AF602044

Aquaculture in Reservoirs

THE AUTHORS

M Menaga has completed her M.F.Sc degree in Aquaculture from Central Institute of Fisheries Education. She won many travel awards to attend International conferences on Asian Pacific Aquaculture. She served as a member of executive expert committee in implementing project on "Sustainable Seafood Industry Development in Myanmar" funded by USAID. She authored five research articles and ten semi scientific articles on advanced aquaculture systems and management practices. She also published many farmer friendly brochures on biofloc technology in shrimp and vannamei farming.

Dr S Felix, Ph.D., Vice Chancellor, Tamil Nadu Dr. J. Jayalalithaa Fisheries University, Nagapattinam is having experience of more than 35 years in the area of fisheries and aquaculture in teaching, research, extension and administration. He has organized more than 20 Nos. of seminar/conference/symposium at National and International levels. Currently he is also the President (2018-19) of World Aquaculture Society, Asian Pacific Chapter and Chairman, ICAR's BSMA Committee (2018-19). He has more than 60 research papers published in National and International journals. He has authored more than 10 books and 25 manuals. He is specialized in the area of advanced systems in aquaculture and aquariculture such as raceway, biofloc technology, RAS, etc,. He has operated around 20 externally funded research projects.

Aquaculture in Reservoirs

by

M Menaga

S Felix

Tamil Nadu Dr. M.G.R. Fisheries College & Research Institute
Tamil Nadu Dr. J. Jayalalithaa Fisheries University
Ponneri, Tiruvallur District, Tamil Nadu

DAYA PUBLISHING HOUSE®
A Division of
ASTRAL INTERNATIONAL PVT. LTD.
New Delhi – 110 002

ISBN: 9789388173186 (Int Edition)

Published by : **Daya Publishing House®**
A Division of
Astral International Pvt. Ltd.
– ISO 9001:2015 Certified Company –
4736/23, Ansari Road, Darya Ganj
New Delhi-110 002
Ph. 011-43549197, 23278134
E-mail: info@astralint.com
Website: www.astralint.com

Laser Typesetting : **Classic Computer Services,** Delhi - 110 035

Printed at : **Neelam Graphics, Delhi - 110007**

डॉ. जे. के. जेना
उप महानिदेशक (मत्स्य विज्ञान)

Dr. J. K. Jena
Deputy Director General (Fisheries Science)

भारतीय कृषि अनुसंधान परिषद
कृषि अनुसंधान भवन-II, पूसा, नई दिल्ली 110 012
INDIAN COUNCIL OF AGRICULTURAL RESEARCH
KRISHI ANUSANDHAN BHAVAN-II, PUSA, NEW DELHI - 110 012

Ph. : 91-11-25846738 (O), Fax : 91-11-25841955
E-mail: ddgfs.icar@gov.in

Foreword

Reservoirs or the 'man-made impoundments' created by a dam of any description on a river, stream or any water course to obstruct the surface flow', primarily serve for irrigation, power generation and other water resource development purposes. Reservoirs in India cover more than one per cent of the country's land surface area. Despite the overwhelming importance of reservoirs with regard to the inland fisheries, the multiple ownership of these water bodies pose a serious bottleneck for overall development of these vast resources.

The production propensity of a reservoir is largely determined by a set of key environmental parameters, especially the water and soil quality, which in turn, are functions of the geo-climatic conditions under which it exists. Thus, the geography, climate, topography and a number of physiographic parameters play a vital role in bestowing the reservoirs their intrinsic productive potential. India, being a country of continental proportions, its reservoirs are spread over various types of terrains and soil types, and exposed to diverse climatic conditions and receiving drainage from a variety of catchment areas.

The present book ***'Aquaculture in Reservoirs'*** aims to disseminate vast potential of Indian reservoirs in the area of culture-based fisheries, especially enabling the students and researchers to have a better access to the information on reservoir ecology, fisheries management and also prospects in aquaculture development. Considering the necessity of greater awareness and understanding on the inter-sectoral and participatory consultations leading to Standard Management Practices, the present book expects to provide needed information for effective management of the reservoirs.

I congratulate and compliment to authors for their sincere efforts in bringing out this useful publication.

(J. K. Jena)

Message

In the developed world, fisheries of inland lakes largely cater to the recreational needs, whereas in a highly populous developing country like India, these resources can play a vital role in augmenting food production for human consumption and mitigating the protein deficiency. India with its vast population and rapidly expanding economy, should invest a scant portion of its economy in research and development studies on potential water resources for aquaculture to provide rational and sustainable use of these resources. Aquaculture is known for its productivity and supply of protein rich food. India's exploitation of inland resources through culture based fisheries is still meagre and it is lowest among countries with active aquaculture sectors. As Indian Reservoirs are referred to as 'Sleeping Giants of the Indian Aquaculture', the exploitation of these resources for the domestic consumption is also unforeseen.

The promotion of Community Based Fisheries in reservoirs is likely to deliver more immediate yield increases than investment in technology in as much as it is technically simpler than conventional aquaculture although there are usually complex technical, social, and institutional issues regarding biodiversity access to water bodies, social equity arrangements and competition of water use to be addressed. This theoretical manual aims to disseminate vast potential of Indian reservoirs in the area of culture based fisheries to enable the students and researchers to have a better access, and to follow up the latest culture technologies. The anticipation of a need for increased supply of fish protein in the sustainable way fashion for the general aspiration expectation of societies can be met by the forward projected ideas. More awareness and understanding, and inter-sectoral and participatory consultations leading to Standard Management Practices, are required for the aquaculture in reservoirs sector development to be effective.

S. Felix
Vice Chancellor
Tamil Nadu Dr. J. Jayalalithaa Fisheries University

Preface

Aquaculture in Reservoirs have been identified as the nearest possible alternative source of fish production recently. India has a 3.25 m ha area of reservoirs and the aquaculture carrying capacity of these reservoirs holds the huge scope in meeting the future fish demand. India with its vast population and rapidly expanding economy, should invest a scant portion of its economy in research and development studies on potential water resources for aquaculture to provide rational and sustainable use of natural resources. The reservoir which is a large man-made ecosystem without a parallel in nature wherein the co-existence of fluviatile as well as lacustrine ecosystem occur, has been conceived as a pillar for development in India.

Neverthless the average fish productivity of reservoirs in India is nearly 30 kg/ha and this is not sufficient enough to match the ever-doubling rate of population demands. The ever-widening gap between the increased demand for fish and the apparent limited growth for natural production from marine capture fisheries has invoked the need for various forms of culture based systems globally. These reservoirs which were constructed for the purpose of generation of hydro-electric power, water supply, irrigation etc. are now being utilized as fishery enhancement units by adopting culture based capture fisheries. This book deals with various fishery enhancement tools that can be used to improve the productivity and production of Indian reservoirs. Any fishery enhancement activity in a reservoir should be based on the morpho-edaphic factors of the reservoir. Keeping this requirement in view, the broad coverage on deployment of cage and pen culture methodologies to enhance fish production and use of advanced aquaculture practices were covered. Limnological profile of the major reservoirs and strategies to improve the fish production in these reservoirs were discussed in vivid . The status and potential for applying fisheries management at small, medium and large reservoirs are reviewed with reference to the revised syllabus based on the fifth Deans Committee and ICAR policies as well as the socioeconomic setting and possible alternatives for management systems.

M. Menaga

S. Felix

Contents

Chapter 1

Nature and Extent of Reservoirs in India

1.1 Reservoir Fisheries Resources of India

Reservoirs, the 'man-made lakes' covering more than 1 per cent of the country's land surface are created primarily for irrigation, power generation and other water resource development purposes. Reservoirs are defined as 'man-made impoundments created by a dam of any description on a river,stream or any water course to obstruct the surface flow.Despite the overwhelming importance of reservoirs in the inland fisheries of India, a reliable estimate of the area under this resource is still elusive, causing serious constraints to the R and D activities. The available estimates made by various agencies are conflicting and wide off the mark. The National Commission on Agriculture (NCA) has estimated the total area under reservoirs at 3 million ha during the mid-sixties and projected its growth to 6 million ha by 2000 AD (Anon., 1976). Bhukaswan (1980) put the figure at 2 million ha. Srivastava *et al.* (1985) compiled a list of 975 large and medium reservoirs in the country with an estimated area of 1.7 million ha. One of its major shortcomings is the exclusion of small reservoirs, especially those in Tamil Nadu, Karnataka and Maharashtra.

1.1.1 Tanks and Small Reservoirs

The word 'tank' is often loosely defined and used in common parlance to describe some of the small irrigation reservoirs in the three southern states of Tamil Nadu, Karnataka and Andhra Pradesh. While in Andhra Pradesh, Karnataka and Tamil Nadu, tanks refer to a variety of water-bodies, some of them are actually irrigation reservoirs, qualifying for inclusion under small and medium-sized reservoirs.

The 'peninsular tanks' is also considered as water bodies created by dams built of rubble,earth stone and masonry work across seasonal streams, as against reservoirs, formed by dams built with precise engineering skills. Enumeration of the medium and large reservoirs is relatively easy, as they are less in number and the details are readily available with the irrigation, power and public works authorities. However, compilation of data on small reservoirs is a tedious task as they are ubiquitous and too numberous to count. The problem is further confounded by ambiguities in the nomenclature adapted by some of the States. The word *tank* is often loosely defined and used in common parlance to describe some of the small irrigation reservoirs. Thus, a large number of small manmade lakes are designated as tanks, thereby precluding them from the estimates of reservoirs. There is no uniform definition for a tank. In the eastern States of Odisha and West Bengal, pond and tank are interchangable expressions, while in Andhra Pradesh, Karnataka and Tamil Nadu, tanks refer to a section of irrigation reservoirs, including small and medium sized water bodies. In fact, some of the tanks in Tamil Nadu and Karnataka are much bigger than Aliyar and Tirumoorthy reservoirs.

David *et al.* (1974) defined the peninsular tanks as *water bodies created by dams built of rubble, earth, stone and masonry work across seasonal streams, as against reservoirs, formed by dams built with precise engineering skill across perennial or long seasonal rivers or streams, using concrete masonry or stone, for power supply, large-scale irrigation or flood control purposes*, which is obviously tedious and inadequate. Irrespective of the purpose for which the lake is created and the level of engineering skill involved in dam construction, both the categories fall under the broad purview of reservoirs, *i.e.*, man-made lakes created by artificial impoundment of surface flow. From limnological and fisheries points of view, the distinction between small reservoirs and tanks seems to be irrelevant. Moreover, numerous small reservoirs fitting exactly into the description of the south Indian tanks are already enlisted as reservoirs in the rest of the country. Therefore, the large tanks have been treated at par with reservoirs for the purpose of this study.

In Andhra Pradesh, the tanks and small reservoirs are segregated either arbitrarily or based on yardsticks that have no limnological relevance. For instance, all the small reservoirs in the State, created before independence and those without a masonry structure and spillway shutters are called tanks. Tanks in Andhra Pradesh are classified as *perennial* and *long seasonal.* Of the 4,604 perennial tanks, 1,804 in Srikakulam, East Godavari and Krishna districts, having average size less than 10 ha, are not considered as reservoirs in this study. The remaining 2,800 tanks covering a total area of 177,749 ha have been reckoned as reservoirs.

In Tamil Nadu, the tanks are classified as *short seasonal* and *long seasonal.* The latter, also known as *major irrigation tanks*, have an average size of 34 ha and retain water for 9 to 12 months a year. Major irrigation tanks of Chengalpattu MGR and Salem districts are larger with average area 222 and 156 ha respectively. A total of 8 837 major irrigation tanks of Tamil Nadu with water surface area of 300 278 ha have been included under small reservoirs. Similarly, 4 605 perennial large water bodies in Karnataka, listed as *major irrigation tanks* are brought under the ambit of reservoirs.

1.1.2 Classification of Reservoirs

Fish Seed Committee of the Government of India (1966) termed all water bodies of more than 200 ha in area as reservoirs. David *et al.* (1974) while classifying the water bodies of Karnataka State, considered impoundments above 500 ha as reservoirs and named the smaller ones as irrigation tanks. In the former USSR, reservoirs up to 10,000 ha area are assigned the status of small reservoirs, whereas in USA they may range from 0.1 ha to several ha (Bennet, 1970). In China, where reservoirs are classified on the basis of storage capacity (Lu, 1986), those holding more than 100 million m^3 of water are classified as large reservoirs, 10 to 100 million m^3 as medium and 0.1 to 10 million m^3 as small reservoirs. Assuming an average depth of 10 m, small reservoirs of China are in the size range 10 to 1,000 ha.

Reservoirs are classified generally as small (<1,000 ha), medium (1,000 to 5,000 ha) and large (> 5,000 ha), especially in the records of the Government of India (Sarma, 1990, Srivastava *et al.,* 1985), which has been followed in this study. All man-made impoundments created by obstructing the surface flow, by erecting a dam of any description, on a river, stream or any water course, have been reckoned as reservoirs. However, water bodies less than 10 ha in area, being too small to be considered as lakes, are excluded.

After removing the anomalies in nomenclature, especially with regard to the small reservoirs, by bringing the large (above 10 ha) irrigation tanks under the fold of reservoirs, India has 19,134 small reservoirs with a total water surface area of 1,485,557 ha (Table 1). Similarly, 180 medium and 56 large reservoirs of the country have an area of 527,541 and 1,140,268 ha respectively. Thus, the country has 19,370 reservoirs covering 3,153,366 ha (Table 2).

The State of Tamil Nadu accounts for maximum number (8,895) and area (315,941 ha) of small reservoirs, followed by Karnataka (4,651 units and 228,657 ha) and Andhra Pradesh (2,898 units and 201,927 ha). Medium reservoirs constitute less than 1 per cent of the total number of units and 17 per cent of the total area. Madhya Pradesh with 169,502 ha tops the table in respect of medium reservoirs. Andhra Pradesh, Rajasthan, and Gujarat have more medium reservoirs than Madhya Pradesh, though the water area in these States is much less. Karnataka has a preponderance in number (12) of large reservoirs. Nevetheless, the 7 reservoirs in Andhra Pradesh in this category are much larger and have a water area of 190,151 ha (Table 2).

The predominance of reservoirs in the peninsular States, *viz.,* Tamil Nadu, Karnataka, Andhra Pradesh, Kerala, Odisha and Maharashtra is elucidated by Figures 1 to 4. These six States account for more than 56 per cent of the total reservoir area in the country. Of the 19,134 small reservoirs, 17,989 (94 per cent) are located there, contributing 63 per cent of the total water area. Similarly, 34 per cent of the medium reservoirs is distributed in these States.

Table 1: Distribution of Small Reservoirs and Irrigation Tanks in India

States	Small Reservoirs		Irrigation Tanks		Total	
	Number	Area (ha)	Number	Area (ha)	Number	Area (ha)
Tamil Nadu	58	15663	8837	300278	8895	315941
Karnataka	46	15253	4605	213404	4651	228657
Andhra Pradesh	98	24178	2800	177749	2898	201927
Gujarat	115	40099	561	44025	676	84124
Uttar Pradesh	40	20845	–	197806	40	218651
Madhya Pradesh	6	172575	–	–	6	172575
Maharashtra	–	–	–	–	–	119515
Bihar	112	12461	–	–	112	12461
Odisha	1433	66047	–	–	1433	66047
Kerala	21	7975	–	–	21	7975
Rajasthan	389	54231	–	–	389	54231
Himachal Pradesh	1	200	–	–	1	200
West Bengal	4	732	–	–	4	732
Haryana	4	282	–	–	4	282
North East	4	1639	–	600	4	2239
Total	2331	551695	16803	933862	19134	1485557

Table 2: Distribution of Small, Medium and Large Reservoirs in India

States	Small		Medium		Large		Total	
	Number	Area (ha)	Number	Area (ha)	Number	Area (ha)	Number	Area (ha)
Tamil Nadu	8895*	315941*	9	19577	2	23222	8906	358740
Karnataka	4651*	228657*	16	29078	12	179556	4679	437291
Madhya Pradesh	6	172575	21	169502	5	118307	32	460384
Andhra Pradesh	2898*	201927*	32	66429	7	190151	2937	458507
Maharashtra	–	119515	–	39181	–	115054	–	273750
Gujarat	676*	84124*	28	57748	7	144358	711	286230
Bihar	112	12461	5	12523	8	71711	125	96695
Odisha	1433	66047	6	12748	3	119403	1442	198198
Kerala	21	7975	8	15500	1	6160	30	29635
Uttar Pradesh	40**	218651	22	44993	4	71196	66	334840
Rajasthan	389	54231	30	49827	4	49386	423	153444
Himachal Pradesh	1	200	–	–	2	41364	3	41564
North-east	4**	2239	2	5835	–	–	6	8074
Haryana	4	282	–	–	–	–	4	282
West Bengal	4	732	1	4600	1	10400	6	15732
Total	19134	1485557	180	527541	56	1140268	19370	3153366

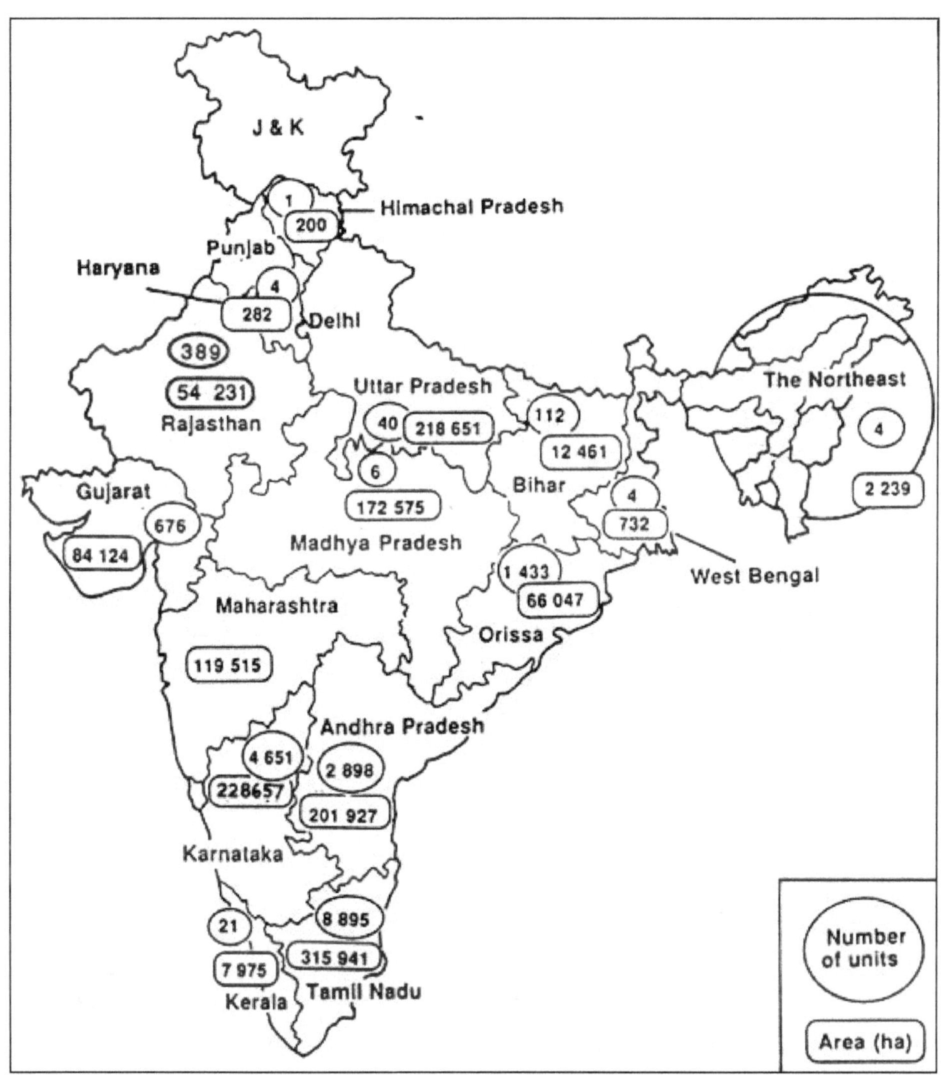

Figure 1: Distribution of Small Reservoirs in India.

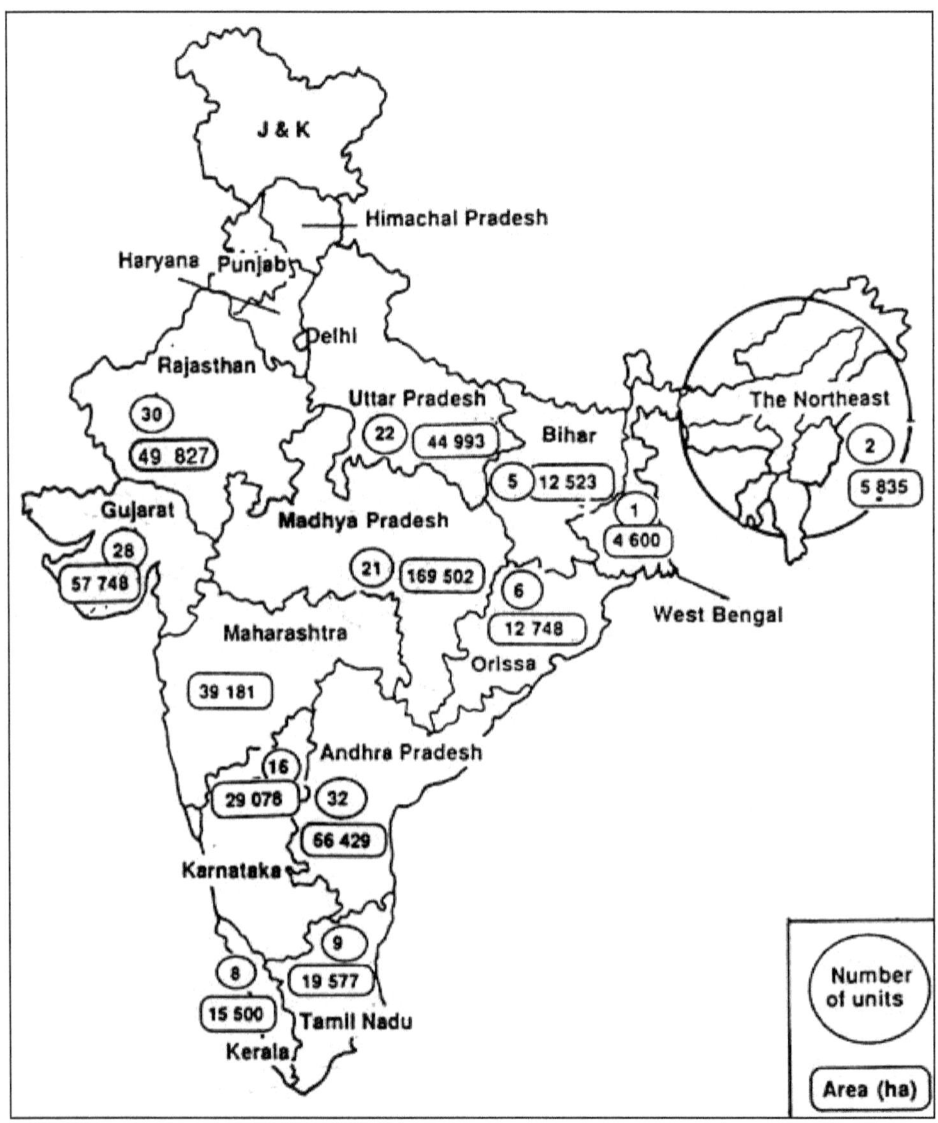

Figure 2: Distribution of Medium Reservoirs in India.

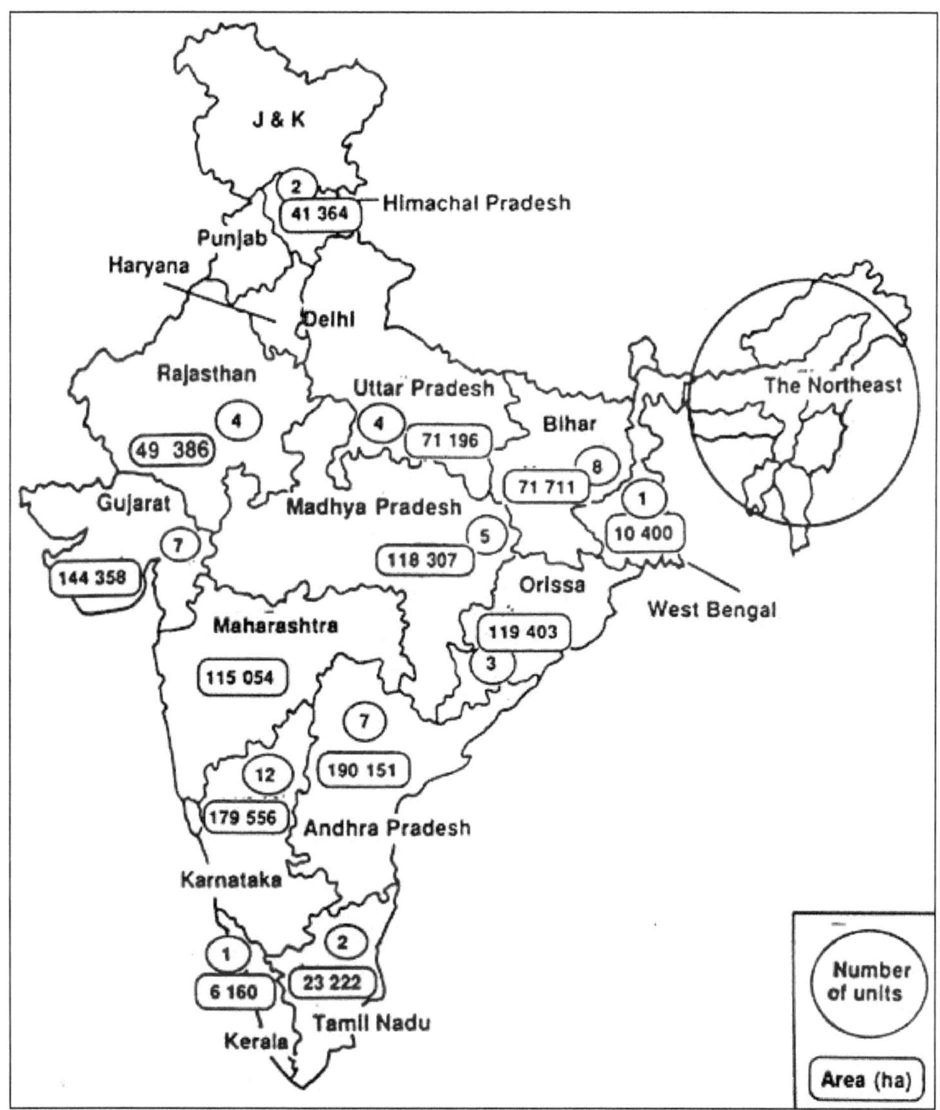

Figure 3: Large Reservoirs in India.

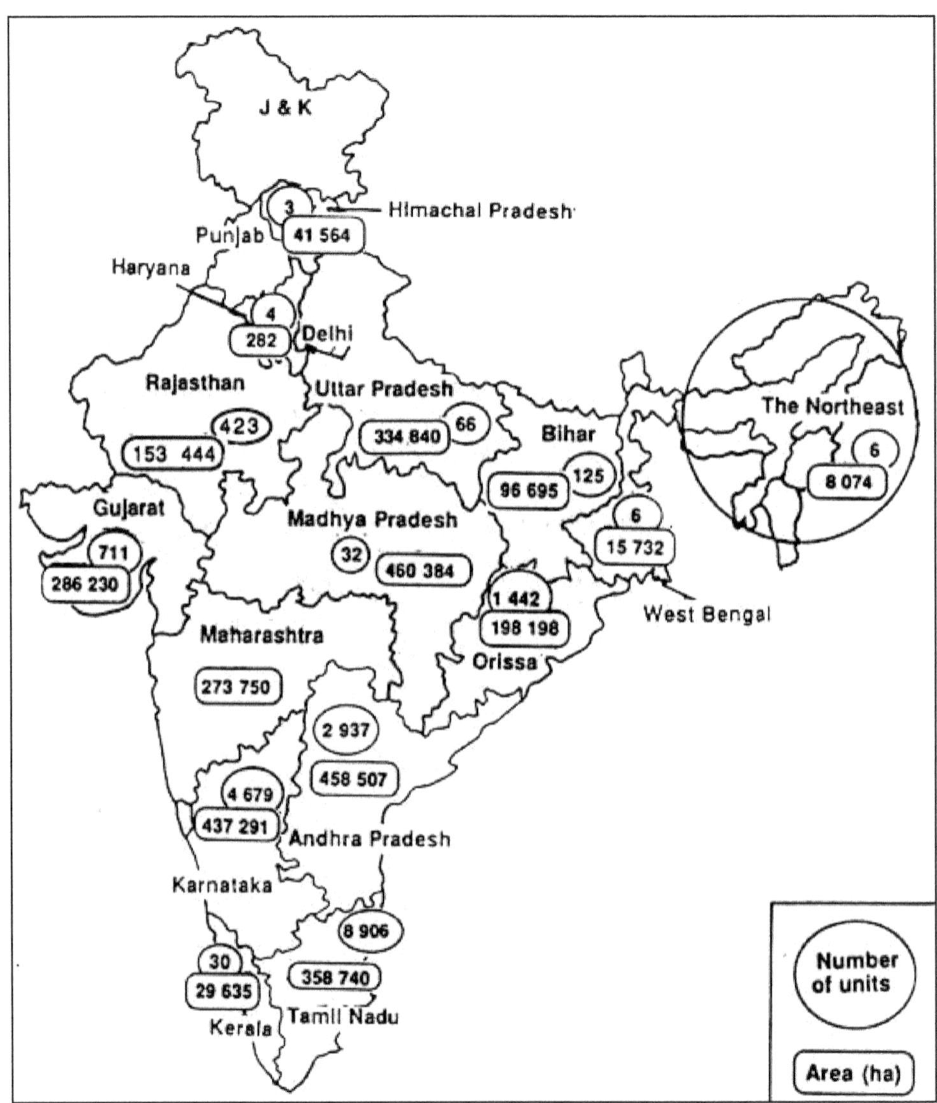

Figure 4: Distribution of Reservoirs (All categories) in India.

1.2 Physico-chemical Features of Reservoirs

Coarse and fine sand predominate the basin soil in the majority of the reservoirs of Rajasthan, Haryana, Uttar Pradesh, and Punjab followed by Madhya Pradesh, Himachal Pradesh, Jharkhand and West Bengal. Silt deposition has been observed mainly along the river course and thus in majority of the reservoirs in Gangetic plains the sediment is mostly sandy-clay type.

1.2.1 Chemical Characteristics

Soil Reaction

Soil is mostly neutral to moderately alkaline in reservoirs reflecting a moderate productive trend. Moderately acidic basin soils were observed in some reservoirs in Bihar and Jharkhand due to the forest-covered catchment. The release of phosphorus is basically dependent on soil pH. Bottom sediment containing a high level of organic matter with slow decomposition rate develops acidity due to humic and short chain fatty acids, leading to less productivity (Shrivastava and Das, 2003). The favourable pH range for biological productivity is 6.5-7.5.

Nutrients

Available Nitrogen

Nitrogen is the basic and primary constituent of protein and is basically present in soil in organic form getting mineralised gradually to ammonium (NH_4^+) and Nitrate (NO_3^-) forms thus making their availability to fish food organisms.

Table 3: Basic Sediment Characters of some Reservoirs (Average values of different case studies)

State	*pH*	*Organic Content (per cent)*	*Available N (mg/100 g)*	*Available P (mg/100 g)*
Himachal Pradesh	6.6-6.7	1.1-1.5	71.-81	5.6.-5.8
Punjab	6.5-7.5	0.25-1.20	45-74	4.8-8.0
Haryana	7.0-7.5	0.5-0.6	32-38	5.8-6.2
Rajasthan	6.8-7.5	0.28-1.20	7-50	0.1-4.0
Uttar Pradesh	6.5-7.6	0.28-3.5	22-60	0.8-2.6
Madhya Pradesh	6.1-8.0	0.4-1.5	25-68	0.7-2.8
Bihar and Jharkhand	5.4-6.5	0.3-0.9	24-47	1.2-8.0
West Bengal	6.4-7.8	0.3-0.8	20-55	0.4-2.4

Reservoirs in Punjab, Himachal Pradesh, Madhya Pradesh, Uttar Pradesh and West Bengal having moderate available nitrogen content but are of moderate to high productivity.

Available Phosphorus

Soil available phosphorus (mg/100 soil) should be in the range of 4.7 to 6.2 for high biological productivity but it is of low profile in Indian reservoirs. Its range (mg/100g) in reservoirs, especially in Punjab, Himachal Pradesh, Jammu and Kashmir, Haryana, was of very high order while it was low in other reservoirs.

Organic Matter and Carbon: Nitrogen Ratio

As per available data comparatively low to moderate organic carbon (per cent) content is recorded in most of the Indian reservoirs. Generally, bottom sediment with 1.5-2.5 per cent organic carbon is considered to be productive.

Nitrogen mineralisation or immobilisation is solely governed by the carbon: nitrogen (C : N) ratio – the higher the C : N ratio (>20) higher the immobilisation of inorganic nitrogen as compared to lower C:N ratio (<10) resulting in mineralisation of organic nitrogen in soil, rendering nitrogen to be added to the water column. Some of the reservoirs in Uttar Pradesh, Bihar and Madhya Pradesh have shown higher C : N ratio due to acidic nature of soil.

1.2.2 Physical Characteristics of Water

Normally the transparency of water decreases during the monsoon due to inflow being loaded with dissolved and suspended organic and inorganic particles, subsequently stabilising in the post-monsoon period. Water transparency has a significant role in light attenuation with depth, affecting productivity. Reservoirs with rocky and gravel catchment such as those in Jammu and Kashmir, Himachal Pradesh and Jharkhand have deeper euphotic zones of more than 2 m (Table 4). In these reservoirs the temporal variation in transparency variation is very wide unlike Madhya Pradesh reservoirs (Unni, 1985). As the reduced transparency in these reservoirs is due to suspended inorganic particles and phytoplankton production is much less, unlike reservoirs with high phytoplankton growth limiting light penetration.

Thermal Stratification

In tropical Indian reservoirs, thermal stratification is not stable and prolonged. The degree of stratification is low in comparison to temperate reservoirs. Even though the period of thermal stratification is short, it has a significant influence on biological productivity. Stable thermal stratification has not been reported in reservoirs of the western, eastern or central parts of India. In a majority of reservoirs, on average, the temperature declined by 1-2°C from surface to bottom. Even the deeper reservoirs such as Bargi, with a maximum depth of 59 m, do not show stable thermal stratification.

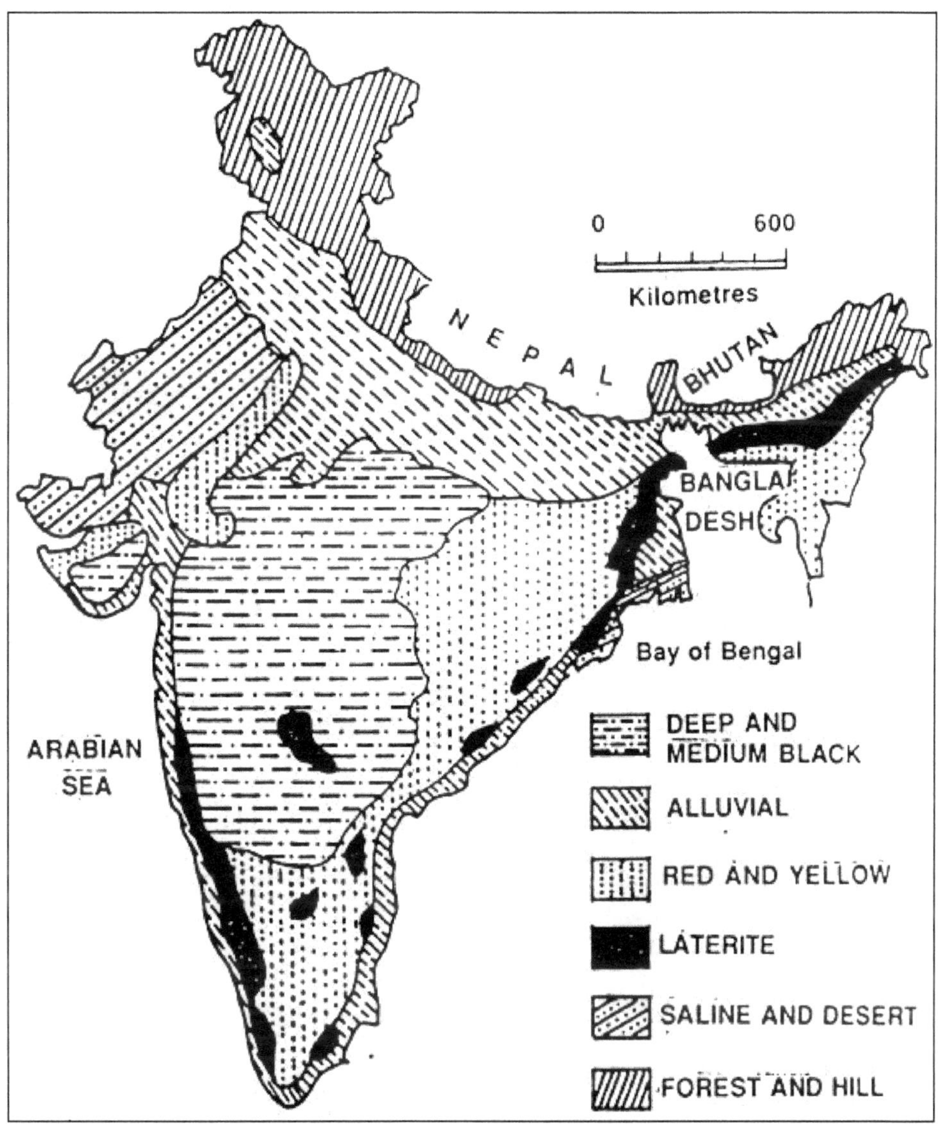

Figure 5: Soil Map of India.

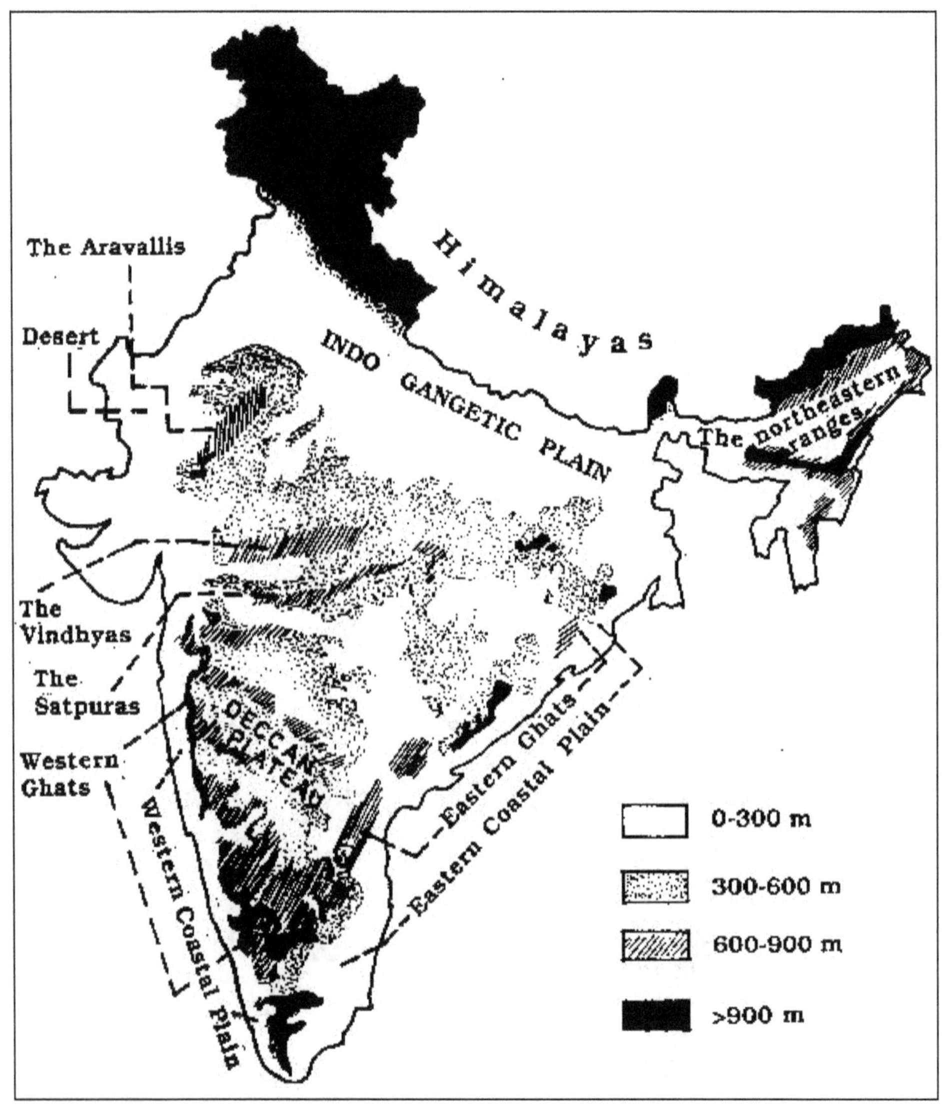

Figure 6: Physiographic Features of India.

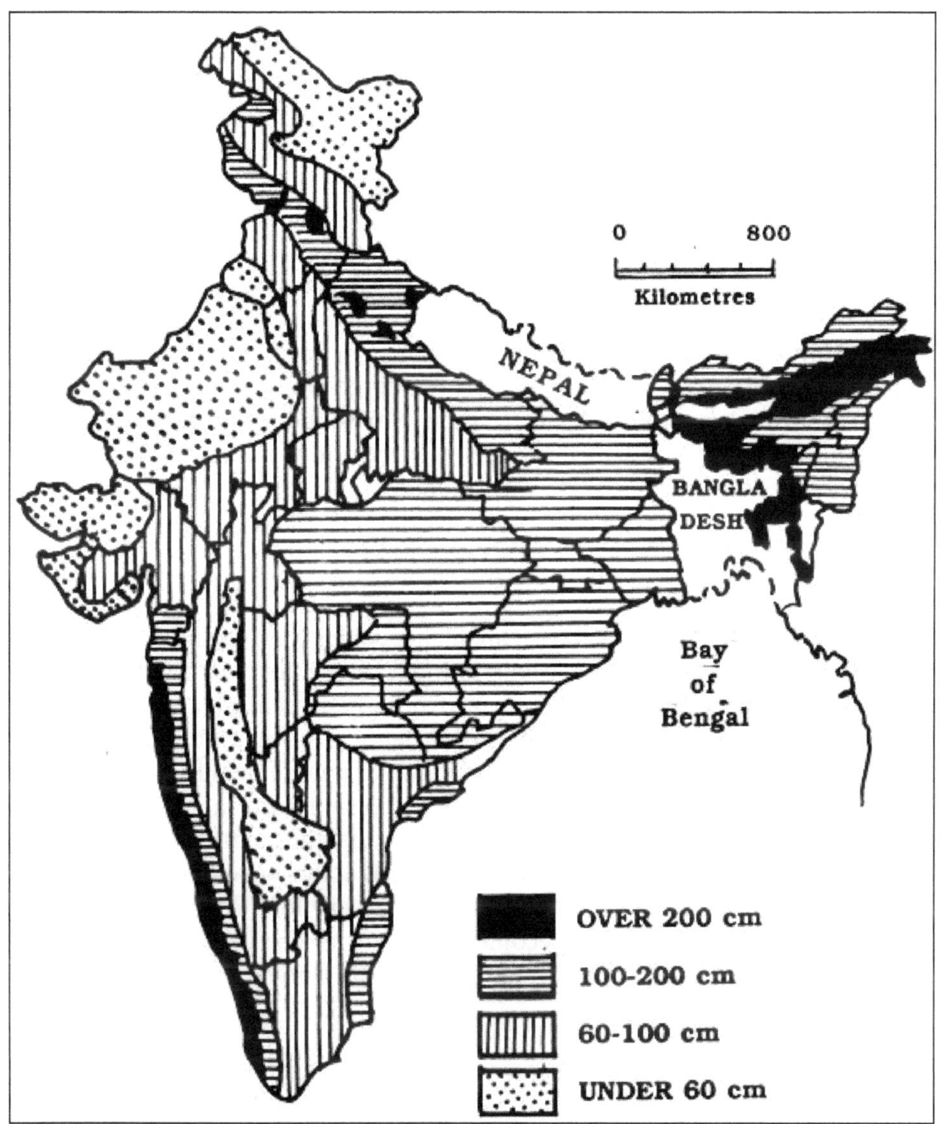

Figure 7: Rainfall Map of India.

In some north Indian reservoirs, thermal stratification occurs during summer with three distinct phases. *e.g.*, in Getalsud reservoir in Bihar (mean depth 4.52m and area 3,400 ha at FRL) the epilimnion extending up to 6m, with a stable thermocline/metalimnion in a range between 7 and 12 m hypolimnion below 12 m (Pal, 1979). The temperature drop in the thermocline region was from 25.3°C at 7m to 20.8°C at 11m. Similar patterns of thermal stratification have also been reported in other north Indian reservoirs such as Konar (Sarakar, 1979) where the metalimnion occurred between 3 to 9 m depth and temperature dropped from 0.7 to 1.1°C per metre. In Rihand (Desai and Singh, 1979), a well defined metalimnion between 4 to 13 m with a fall of temperature between surface and bottom at 10°C, 8.5°C and 9.5°C in May 1975, 1976, 1976 and 1977, respectively has been reported. In Govindsagar reservoir, a very strong and well defined thermocline was observed (Sarkar *et al.*, 1977). Thermal stratification with a fall in metalimnion temperature of less than 1°C has been observed in many tropical lakes (Le-Cren *et al.*, 1980). Similar trends are evident in the upper peninsular reservoir comprising South Bihar, Madhya Pradesh and Gujarat which undergo transient phases of thermal stratification during summer. Apart from the year-round high-water temperature and continuous draw down from deeper layers, the wind and wave mediated turbulence facilitate mixing of the water column.

1.2.3 Chemical Features of Water

Dissolved Oxygen

Generally, dissolved oxygen (DO) are higher during the pre- and post-monsoon seasons than in the monsoon periods in India reservoirs due to the downpour. DO is greater during March-April in some tropical and peninsular reservoirs due to *Microcystis* blooms. An oxycline is a basic feature in any productive reservoir. Deep tropical and sub-tropical reservoirs show a stable clinograde (*i.e.*, decreasing DO levels with depth) with sharp fall in oxygen concentration at the hypolimnion, which mostly occurs during pre-monsoon. The longer the duration of such stability the more productive is the reservoir. Highly productive water should have a DO concentration of more than 5mg/l. However, a very high concentration of DO leading to super saturation may become lethal to fish fry, as has been seen in some reservoirs predominantly infested with Microcystis blooms.

Free Carbon Dioxide

In some of the reservoirs, free from macrophyte vegetation or being confined to limited littoral zone only, free CO_2 was in trace or absent at surface waters. In contrast, a substantial amount of free CO_2 was in trace or absent at surface waters in the major reservoir of Madhya Pradesh in the range from 2.0 to 42.0 mg/l in Bargi, Tawa and Halali (Unni, 1993) and 16.0 to 30.0 mg/l in Rihand (singh *et al.*, 1980).

pH

Due to natural buffering capacity of water, Indian reservoirs seldom show drastic pH fluctuations. However in general, water pH is low during the monsoon due to dilution of alkaline substances or dissolution of atmospheric CO_2 with a resultant increase in the post-monsoon and pre-monsoon periods. The majority of the reservoirs have moderately alkaline pH due to alkaline soil reaction.

Specific Conductivity

Specific conductivity is an index of the amount of water soluble salts present in water. Barring West Bengal and Jharkhand, most of the reservoirs in other states recorded conductivity in the range 100 to 300 mS/cm with some exceptionally higher values in case of reservoirs in Rajasthan and Haryana (Table 4). In normal freshwater, chloride content lies within 10 to 30 mg/l; if the value is somewhat more that this, it indicates local pollution. The average chloride content (mg/l) in reservoirs was in the range 7-30 with higher values observed during the pre-monsoon months in some reservoirs of Rajasthan and Haryana (Table 4). In freshwaters, electrical conductivity is also considered to be an index of productivity when the ionic composition is dominated by CO3- and HCO3 system because conductivity is known to be related to total dissolved solids (Henderson and Welcomme, 1974).

Alkalinity

The majority of reservoirs have shown moderate total alkalinity within the range of 40-205 mg/l representing moderate productivity trends (Table 4). The low values of total alkalinity in Jharkhand and Bihar reservoirs are due to the acidic basin soil of lateritic type originating from dry forest areas; the dominance of carbonate is least pronounced in these reservoirs. In general, total alkalinity is high during pre-monsoon with a substantial decrease in the monsoon period due to dilution.

Table 4: Ranges of Physico-chemical Characteristics of some Reservoirs in different Stages

States	*Transparency (m)*	*Alkalinity (ppm)*	*Conductivity (μ/CM)*	*Chloride (ppm)*
Jammu & Kashmir	0.1-0.8	46-112		
Himachal Pradesh	115-180	46-84	95-166	11-14
Punjab	0.5-1.0	73-198	168-300	12-15
Haryana	0.4-1.0	114-205	213-764	27-86
Rajasthan	0.3-1.5	40-256	142-959	7-70
Uttar Pradesh	0.2-3.4	44-164	72-280	3.5-25.0
Madhya Pradesh	0.2-1.5	81-170	124-337	14.5-44.5
Bihar & Jharkhand	0.1-1.1	43-63	12-306	14-65
West Bengal	0.1-1.2	40-125	28-154	12-34

1.2.4 Nutrient Status

In Indian reservoirs, available-N (NO_3 + NO_2) in water is typically very low (Table 5) throughout most of the year except in monsoon and to some extent for a shorter period during the pre-monsoon period. In the monsoon its availability is due to run-off from fertile catchments and in summer it is due to a faster rate of decomposition of bottom organic load, rendering nitrogen availability to standing water phase. The peak values of nitrate-N are encountered in the monsoon months and to a lesser extent in pre-monsoon periods.

In Indian reservoirs, except for a shorter period during the monsoon, the availability of phosphate is of a very low order and rarely exceeds 0.1 mg/l (Table 5). The lack of these nutrients in water does not seem to be indicative low productivity, particularly in reservoirs free from pollution because of their rapid turnover and quick recycling *e.g.* 95 per cent of the phosphorus is taken up by phytoplankton within 20 minutes while some algae can convert inorganic phosphate into more insoluble organic state within one minute (Hayes and Philips, 1958).

Table 5: Ranges of Basic Nutrients in Water of some Reservoirs in different States

State	*Nitrate-N (ppb)*	*Phosphate-P (ppb)*	*Silicate-Si (ppm)*
Himachal Pradesh	10-250	40-150	3.0-5.5
Punjab	10-300	10-200	2.5-4.4
Haryana	10-150	5-200	1.6-3.8
Rajasthan	5-2020	10-280	0.27-11.21
Uttar Pradesh	10-430	20-650	3.1-9.5
Madhya Pradesh	5-469	5-180	0.8-8.5
Bihar & Jharkhand	5-600	2-330	1.3-33.0
West Bengal	5-96	0-110	2.2-7.4

1.2.5 Biological Features of Reservoirs

Plankton takes the lead role in maintaining trophic structure as either the macrophytes nor the periphyton get a strong hold due to draw down effects and high flushing rates in Indian reservoirs. The plankton trends in some reservoirs are given in Table 6.

In most of the reservoirs studied, green algae and blue-green algae from the bulk of the plankton community followed by diatoms. The occurrence of *Microcystis* is significant. Among zooplankton, copepods are important followed by *Cladocerans*, rotifers and protozoans. Altitude plays an important role in the distribution of Microcystis. In Gobindsagar, which is a productive sub-temperate reservoir, *Ceratium* sp. Was dominant over *Microcystis.* The Rajasthan reservoirs which receive scanty rainfall and have poor flushing rates favour macrophytes and blooms of *Microcystis* are not recorded. Plankton usually exhibits two peaks in a year; a distinct winter pulse attributed to nutrient rich higher monsoonal inflow and summer pulse of lower magnitude, Monsoonal pulses have also been observed in some reservoirs.

Macro-benthos

According to the studies reported (Katiha *et al.,* 2007) shallow reservoirs such as Bachhra, Baghla, Kerwa and Dahod were found to be particularly which in benthic fauna due to favourable substratum, rich organic matter and minimal water level fluctuations. The deeper reservoirs like Rihand had poor benthic macrofauna whereas the reservoirs like Chamera with a rocky and gravely bottom had no microbenthic communities. Dipterans were most important in the reservoirs followed by gastropods, bivalves, annelids and other oligochaetes. The importance of bottom fauna in relation to fish yield in reservoirs has been highlighted by hayes (1957).

Table 6: Plankton Trends in some Reservoirs of different States

State	*Phytoplankton*	*Zooplankton*	*Plankton*
Himachal Pradesh	80	20	114-198
Punjab	58-87	12-42	319-435
Haryana	75-76	23-24	490-664
Rajasthan	70-85	15-30	583-4503
Uttar Pradesh	70-85	15-30	250-1500
Madhya Pradesh	60-94	6-40	150-3550
Bihar & Jharkhand	67-94	6-33	200-3000
West Bengal	70-80	20-30	150-500

Macrophytes

Some reservoirs have a good growth of macrophytes due to low flushing rates, *e.g.,* Ramgarh, Jaisamand (Rajasthan), Bachhra, Baghla, Pahuj (Uttar Pradesh), Halali and Dahod (Madhya Pradesh). The biomass has varied from 0.2 to 5.2 kg m^{-2} wet weight. The peak is vidually observed in summer and decreased during the monsoon season. Macrophytes were mostly submerged, emergent and floating types and were always dominant in the littoral areas. Macrophytes provide substrata to many insects, molluscs and other invertebrates. They accumulate large quantities of inorganic nutrients affecting phytoplankton growth adversely. The submerged weeds give shelter to minnows, which compete with major carps for food. Heavy growth of macrophytes prevents light penetration and cases of fish mortality due to heavy macrophyte growth have been recorded from Rajasthan. They also pose serious problems in the operation of fishing gear.

Among the macrophytes *Potamogeton* was predominant while *Hydrilla, Vallisneria, Ceratophyllum* and *Najas* were also recorded. These are generally impediments for operation of fishing nets but during monsoon, these submerged macrophytes facilitate the collection and anchoring of fish eggs.

Primary Productivity

The gross primary production (GPP) in the reservoirs was in the range of 25-950 mg C m^{-3} hr^{-1} (Table 7) with wide temporal variations. Assimilation efficiencies were very moderate (55.40-74.30 per cent) signifying that these water bodies are moderately productive. The estimated fisheries potential based on GPP varied from 45-545 kg $ha^{-1}yr^{-1}$ in different states. The existing fish yield was very low as compared to the potential. Based on primary production trends Natrajan and Pathak (1983) worked out energy flow estimates in some reservoirs in India, which gives an approximate estimate of potential food chain behaviour in the ecosystem.

Table 7: Ranges of Gross Primary Production (GPP) in some Reservoirs in different States

States	*GPP (mg C m^{-3} day)$^{-1}$*
Himachal Pradesh	1,008-1,560
Punjab	2,016-2,784
Rajasthan	1,440-3,936
Uttar Pradesh	600-2,640
Madhya Pradesh	936-22,800

Fish Production Trends

In India, reservoirs are under-exploited with an average annual fish yield of 20.15 kg/ha, with mean production in small, medium and large reservoirs being 48.05 kg/ha/year, 12.39 kg/ha and 11.31 kg/ha respectively. In the case of small reservoirs, Madhya Pradesh (47.26 kg/ha/year) and Rajasthan (46.43 kg/ha/year are important states in terms of fish production. The contribution of medium reservoirs to reservoir fish production is comparatively higher from Rajasthan (24.47 kg/ha/year). The fish yield of large reservoirs in Madhya Pradesh (40 kg/ha/year) and Himachal Pradesh (35.55 kg/ha/year) is high as compared to the reservoirs of other states. However, the overall fish yields from reservoirs remained very low (Table 8). This is mainly due to poor management as a result of the lack of understanding of reservoir ecology, trophic dynamics. Inadequate stocking, wrong selection of species for stocking, low size of stocking materials and irrational exploitation. Adopting the available package of practices, developing the necessary infrastructure facilities and providing conducive socio-economic environment, can bridge the existing wide gap between the potential and the actual yield.

Based on hydrological condition and primary productivity, the large reservoirs have a production potential of 65-190 kg/ha, the medium reservoirs 145-215 kg/ha and small reservoirs 285-545 kg/ha per year. There is a wide gap between potential and the actual yield, which can be mitigated achieving 70-80 per cent of the potential through proper scientific management. The fish yield of mot of the Gangetic Basin

Table 8: Average Annual Fish Yields from Reservoirs in different States of India (kg/ha)

State	Small	Medium	Large	Pooled
Tamil Nadu	48.5	13.7	12.7	22.6
Uttar Pradesh	14.6	7.2	1.1	4.7
Andhra Pradesh	188.0	22.0	16.8	36.5
Maharashtra	21.1	11.8	9.3	10.2
Rajasthan	46.4	24.5	5.3	24.9
Kerala	53.5	4.8	–	23.4
Bihar	3.9	1.9	0.1	0.1
Madhya Pradesh	47.3	12.0	14.5	13.7
Himachal Pradesh	–	–	35.7	35.6
Odisha	25.9	12.8	7.6	9.7
Average	49.5	12.3	11.4	20.1

Table 9: Average Annual Fish Yields in Reservoirs in different States in India

State	Yield (kg/ha)			
	Small	Medium	Large	Total
Madhya Pradesh	47.3	12.0	14.5	13.7
Bihar	3.9	1.9	0.1	0.1
Uttar Pradesh	14.6	7.2	1.1	4.7
Rajasthan	46.4	24.5	5.3	24.9
Himachal Pradesh	-	-	35.6	35.6
Total for IGB states	29.8	12.8	9.2	18.3
Total India	49.9	12.3	11.4	20.1

Table 10: Production Potential of Small Reservoirs in India (kg/ha/year)

Reservoirs (ha)	Potential	Actual
Manchanbele (329)	694	118
Aliyar (320)	250	193.8
Thirumurthy (388)	268	213
Sarni (1012)	140	56
Nelligudda (80)	840	1825
Kyrdemkulai (80)	571	-
Gulariya (300)	200	150
Bacchra (140)	240	139
Baghla (250)	200	104
Govindgrah (307)	220	60
Kulgarchi (193)	45	7.4
Loni (350)	36	27

reservoirs was less than overall average of the country, barring medium reservoirs, which was marginally higher. Overall the fish yield was maximum in the state of Himachal followed by Rajasthan and Madhya Pradesh (Table 9). Considering the existing and potential fish productivity, reservoirs have very good scope for increasing fish production and productivity (Table 10).

Chapter 2

Trophic Phases in Reservoir

2.1 Reservoir Ecosystem

Reservoir is a man-made ecosystem without a parallel in nature. Though essentially a combination of fluviatile and lacustrine systems, a close examination of the biotope reveals that it has certain characteristic features of its own. The riverine and lacustrine characters coexist in reservoirs, depending on the temporal and spatial variations of certain habitat variables. For example, the lotic sector of the reservoir sustains a fluviatile biocoenos, whereas the lentic zone and the bays harbour lentic communities. During the months of heavy inflow and outflow, the whole reservoir mimics a lotic environment whereas in summer, when the inflow into and outflow from the reservoir dwindle, a more or less lentic condition prevails in most parts of it. Another unique feature of reservoirs that makes them distinctly different from their natural counterparts is the water renewal pattern marked by swift changes in levels, inflow and outflow.

In India, most of the precipitation takes place during the monsoon months which contribute substantially to the surface flow. During this period, due to a heavy inflow of water into the system, all outlets of the dam are usually opened, resulting in total flushing. This process dislodges a considerable part of the standing crop of biotic communities at the lower trophic level and disturbs the natural primary community succession. The sudden level fluctuations also affect the benthos by exposing or submerging the substrata. Factors determining the water and soil quality in reservoirs are different from those of natural lakes. In the latter, the basin soil plays a predominant role in determining the chemical water quality through soil water interphase. In the reservoirs, on the other hand, the nutrient input from the allochthonous source often determines the water quality, nutrient regime and the basic production potential. This is because of the fact that the catchment of parent rivers is very often situated far away from the reservoir, under totally different geoclimatic conditions. Deep drawdown, wind-mediated turbulence, locking up of nutrients in the deep basins, *etc.*, are some of the factors that impart the uniqueness to the reservoir ecosystem. Besides, the varying purpose and design of the dams make the reservoirs different in their hydrographic and morphoedaphic characteristics, with implications on the production potential. Some salient features of the reservoir ecosystem are depicted in Table 11.

Table 11: Abiotic and Biotic Factors Affecting Productivity of Reservoirs at Various Trophic Levels

Positive/Augmentative Factors	*Major Effects unknown*	*Negative/Reductive Factors*
High shoreline development (coves, bays, bays *etc.*)	Sedimentation of inorganic material	Low transparency in floods due to
Low mean depth (less than 18 m)	High rate of evaporation	High mean depth
Existence and extent of marginal vegetation	Contributions of autochthonous nutrients	Erosion in the reservoir watershed area
Optimum nutrient levels	High surface temperature during summers (in northem India)	Reduction of quantity of water flowing into reservoir
Nutrient enrichment during floods	Low water temperature during winter (In northern India)	Large water level fluctuations creating large aridal (barren littoral)
Moderate to long growing season	Aquatic community interrelationship	Low level of dissolved oxygen in parts of hypolimnion
High frequency of phytoplankton blooms	–	Pollution in the reservoir watershed
Moderately developed macrophyte community	–	Phytoplankton biomass mainly blue greens
Periphyton abundant	–	Relative low fish species diversity indicating low stability and a potentially low resilience to stresses
Well established plankton and benthos	–	Unbalanced fish populations favouring predatory and trash species
Tree and bush cleared	–	Low abundance and diversity of terrestrial vegetation hence early successional stage
Conditions permitting passage of migratory fish	–	Relatively low environmental heterogeneity
Introduction of fishes adapted to lentic conditions	–	Low diversity of plankton and benthos
Employment of modern fishing gear and optimization of fishing effort	–	Low diversity of aquatic macro-phytes
Enforcement of fishery regulations	–	Exposure of fish nests during draw-downs

(After Jhingran, 1988).

The ecology of reservoirs is radically different from that of the parent rivers. Dams alter river hydrology both up- and downstream of the river. The obstruction of river flow and the consequent inundation trigger off sudden transformation of lotic environment into a lentic one. The riverine community is subjected to changes akin to the secondary community succession. A number of organisms perish, some migrate to more hospitable environs, and the more hardy ones adapt themselves to the changed habitat. There is usually an initial spurt of plankton and benthic communities due to the increased availability of nutrients released from the decay of submerged vegetation. This trophic burst is also on account of the saproxenic lacustrine species filling the vacant niches created by the disappearance of saprophobic riverine taxa. As the effects of trophic burst wean away, the reservoir passes into a phase of trophic depression and the final fertility is regained after a few years. Habitat variables responsible for a reservoir's productivity can be summed up into climatic, morphometric, and hydro-edaphic factors.

2.2 Determinants of Reservoir Productivity

The production propensity of a reservoir is determined by a set of key environmental parameters, especially the water and soil quality which, in turn, are functions of the geo-climatic conditions under which it exists. Thus, the geography, climate, topography and a number of physiographic parameters play a vital role in bestowing the reservoirs their intrinsic productive potential. India, being a country of continental proportions, its reservoirs are spread over various types of terrains, and soil types exposed to diverse climatic conditions and they receive drainage from a variety of catchment areas.

2.2.1 The Geoclimatic Features

The land area of India covers 3 287 728 km^2, half of which lying above the Tropic of Cancer and the rest in the tropics. The southern limit is as close to equator as 8° 4' N. The climate varies from the warm tropical in the south to the temperate in the north. The landscapes include some great mountains, extensive alluvial plains, riverine wetlands, plateau lands, deserts, coastal plains and deltas. The main soil types are alluvial, deep and medium black, red and yellow, laterite, saline and desert, and forest and hill (Figure 5). Almost all conceivable forms of vegetation, including tropical evergreen, littoral and swamp, tropical moist deciduous, tropical thorn, montane sub-tropical, Himalayan and alpine are present in various parts of the country.

The major physiographic divisions of the country are the Himalayas, the Indo-Gangetic plains, the Vindhyas, the Satpuras, the Western Ghats, the Eastern Ghats, coastal plains, the deltas and the riverine wetlands (Figure 6). The alignments of hills and their elevation have profound influence on the prevailing winds and thereby the distribution of rainfall in the country. India receives, on an average, 10^5 cm of rainfall every year, which is one of the highest in the world for a country of comparable size. Total amount of rainfall received annually is estimated at 400 million hectare meters (mhm), out of which 230 mhm goes back to the atmosphere as evapotranspiration, leaving 170 mhm to impregnate the rivers through surface flow (110 mhm) and regeneration (60 mhm). The temporal and spatial distribution of rainfall exhibits wide variations within the country.

More than one million km^2 of the country's geographical area receives inadequate rainfall. This includes the deserts, the semi-arid regions of north India and the rain shadows of the Western Ghats (Figure 7). Large rivers like the Godavari, the Krishna, the Pennar and the Cauvery pass through extensive tracts of low rainfall areas and hence carry much less water than the rivers passing through high rainfall areas like the north-east and the west coast. In the north-east, the Khasi and Jaintia hills receive a bountiful 1,000 cm of rainfall annually and the Brahmaputra valley gets precipitation to the tune of 200 cm. Rainfall up to 1,142 cm recorded in Cherrapunji and Mawsyngram in the region is one of the highest in the world. In the west coast of India, heavy rainfall occurs along the slopes of the Western Ghats, where during the southwest monsoon, rainfall of very high order is recorded on the windward side which rapidly decreases towards the leeward side. The Indo-Gangetic plains and the Himalayas also receive rainfall above the national average.

Seasonal distribution of rainfall in India is worth noticing. The Western Ghats, Assam, parts of sub-Himalayan West Bengal and some higher elevations of Himalayas up to Punjab have more than 100 rainy days a year, while in extreme west Rajasthan the number of rainy days are less than 10 (Rao, 1979). The south-west monsoon season extending from June to September is the principal rainy season in the country as a whole when 75 per cent of the annual rainfall is received. In more than one third of the country, 90 per cent of the rainfall and thereby the surface flow is limited to a very brief period of 2 to 3 months. This extreme seasonality in rainfall distribution makes the irrigation reservoirs a *sine qua non* for agriculture in India, especially in the rainshadows of the peninsular India. People inhabiting this area learned to store water by erecting barricades across minor stream and rivulets from time immemorial. In recent years, with the advent of modern hydraulic structures, larger and more complex dams came into being. The steep gradient and heavy discharge of water in the mountain slopes of Western Ghats, the north-east and the Himalayas offer ideal opportunities for hydro-electric power generation. A large number of such projects have come up in these regions in recent years. Thus, the reservoirs have become a common feature in the Indian landscape, dotting all river basins, minor drainages and seasonal streams.

2.2.2 Climatic Factors

The Indian reservoirs are exposed to a wide range of climates from the temperate Himalayas in the north to the extreme tropical in the southern peninsula. From Gobindsagar in Himachal Pradesh to Chittar in Tamil Nadu, they spread over the southern slopes of Himalayas, the Indo-Gangetic plain, the Vindhyas, the Satpuras, the Western and Eastern Ghats and the Deccan plateau.

Apart from being the main factor influencing the prevailing climate of the region, the latitudinal location is important in determining the quantum of solar radiation available at the water surface for primary productivity. Natarajan and Pathak (1983) estimated the amount of solar radiation available at four reservoirs within 31° 25′ N and 11° 28′ N and the rate at which the solar energy was converted into chemical energy. Incident solar energy available at the surface varied from 213×10^4 cal m^{-2} yr^{-1} in Bhavanisagar (10°28′ N) to 172×10^4 cal m^{-2} y^{-1} in Gobindsagar (31°25′ N). Jhingran (1990) observed that 0.2 to 0.68 per cent of the incident solar energy was fixed as chemical energy by the primary producers in five reservoirs, *viz.*, Gobindsagar (Himachal Pradesh), Ramgarh (Rajasthan). Rihand (Uttar Pradesh) and Bhavanisagar (Tamil Nadu). It is often the qualitative and quantitative abundance of the producer communities that determines the photosynthetic efficiency rather than the actual amount of solar energy available.

For instance, Nagarjunasagar despite receiving solar energy at the rate of 205 $\times 10^4$ cal m^{-2} yr^{-1}, fixes chemical energy to the extent of 0.29 per cent, whereas in Gobindsagar 0.68 per cent of the 172×10^4 cal m^{-2} yr^{-1} is being fixed by the producers.

Prevailing climatic factors including air temperature, wind velocity, rainfall, *etc.* play an important role in the biological productivity of a water body. The wide seasonal variations in air temperature is the predisposing factor in the thermal features of the north Indian and peninsular reservoirs. In contrast to the reservoirs of the north, their southern counterparts are characterised by the narrow range of fluctuations in water and air temperature during different seasons, a phenomenon which prevents the formation of thermal stratification. Normally, the thermal gradient occurs when high air temperature during summer warms up the upper layer. In peninsular India, there is no winter worth its name and the air temperature remains comparatively high during the whole year. During summer, when surface

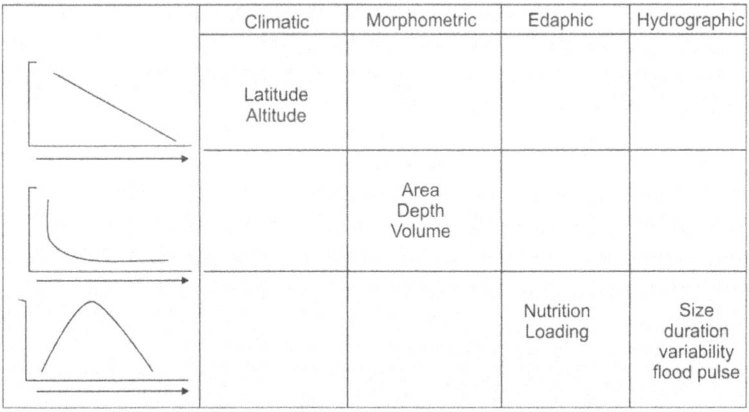

	Climatic	Morphometric	Edaphic	Hydrographic
	Latitude Altitude			
		Area Depth Volume		
			Nutrition Loading	Size duration variability flood pulse

Figure 8: Generalized Effects of Climatic, Morphological, Edaphic and Hydrological Factors on Lake Productivity.

water gets heated up, the prevailing high temperature at the bottom does not offer any scope for thermal resistance by the warm upper layers. Thermal stratification is limnologically important because in thermally stratified lakes, the water above and below thermocline does not mix up and thereby rich nutrients at the bottom layer get locked up. A warm bottom layer also facilitates rapid decomposition of organic matter, thereby accelerating the process of nutrient release.

Table 12: Latitude-Induced Variations in Availability of Energy in Indian Reservoirs

Reservoir	*Area (ha) at FRL*	*Latitude*	*Available Light Energy (cal m^{-2} yr^{1})*	*Available Chemical Energy (cal $m^{-2}yr^{1}$)*
Bhavanisagar	7,285	11°25'	213×10^4	8,781 (0.41 per cent)
Nagarjunasagar	2,8474	16°4'	205×10^4	5,959 (0.29 per cent)
Rihand	4,6538	24°	188×10^4	3,970 (0.20 per cent)
Ramgarh	1,265	27°12'	183×10^4	8,236 (0.49 per cent)
Gobindsagar	16,867	31°25'	172×10^4	11,696 (0.68 per cent)

After Jhingran, 1990.

The deep basin of Nagarjunasagar, despite 40 per cent of its capacity being dead storage, does not favour the formation of thermocline. Apart from the high water temperature throughout the year, the continous drawdown from deeper layers and the wind and wave mediated turbulence facilitate mixing up of water column. This is true to most of the reservoirs in the States of Andhra Pradesh, Karnataka, Kerala, Tamil Nadu, Odisha and Maharashtra. However, seven reservoirs in the upper peninsula comprising south Bihar, Gujarat and Madhya Pradesh undergo transient phases of thermal stratification during the summer stagnation, depending upon the other parameters such as the depth of the basin, water abstraction pattern and the wind. Konar reservoir situated beyond the Tropic of Cancer has distinct epi-and hypolimnion during the summer months. Similarly, a well-defined thermocline is reported from Gobindsagar. Apart from the solar warming of the top layer, which remains as a separate thermal regime, the inflowing Beas water that joins the reservoir at the lotic sector does not get mixed up, retains its cool character and remains as a separate layer at the bottom.

The amount of rainfall determines the rate of inflow into the reservoir, and hence plays a crucial role in bringing in the water replenishment and nutrient enrichment. More often, rainfall in the catchment of the river situated hundreds of km away from the reservoir affects the inflow rate. Another important climatic factor with implications on thermal and chemical regimes of the reservoir is the wind. It helps distribution of heat and equalisation of temperature in the water column. Wind velocity is very high in monsoon and premonsoon months in most reservoirs in India (Natarajan, 1979). Wind-induced turbulence is important in churning of the reservoirs and thereby facilitating the availability of nutrients at the trophogenic zone.

2.2.3 Morphometric Factors

Reservoir morphometry is a function of the height of the dam and the topography of the impounded areas. Apart from the nature of the basin and the characteristics of the terrain, it is the design of the dam and the water use pattern that decide the influence of morphometric and hydrographic features on the aquatic productivity. Most of the hydel reservoirs on the mountain slopes of Western Ghats, Himalayas and the other highlands are deep, with steeper basin walls than the irrigation impoundments.

One of the most important morphometric considerations is the mean depth, that is believed to determine the productivity of reservoirs (Hayes, 1957, Rawson, 1952). This is based on the well known dictum that the shallower lakes have greater part of their water in the euphotic zone, facilitating greater mixing and circulation of heat and nutrients and hence higher productivity. A large portion of water in deep lakes serves as a *nutrient sink* at the bottom, where organic matter accumulates and thus the nutrients become unavailable at the photosynthetic zone. Mean depth calculated from the capacity and area varies from 5.2 in Panchet to 58 m in Gobindsagar among the large reservoirs. Medium reservoirs in the country have mean depth ranging from 2.3 m (Poondi) to 24.0 m (Bhatghar). Hope Lake in Tamil Nadu has an exceptionally deep basin of 37.7 m, while the small reservoirs in India have a mean depth range of 2.1 m (Vidur) to 14.57 (Badua). Mean depth in case of hydel reservoirs of the mountain slopes is invariably high, compared to the irrigation reservoirs of the plains and plateaus. The two largest impoundments in

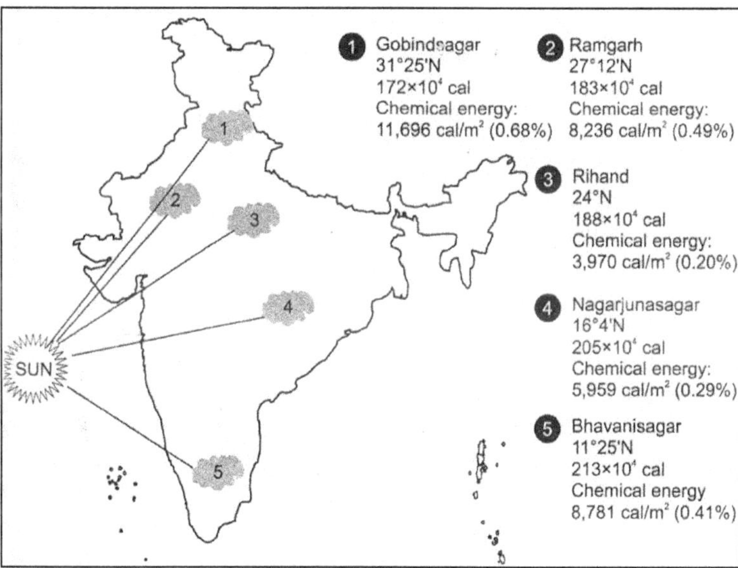

Figure 9: Effect of Latitudinal Location in Primary Productivity of Reservoirs.

the country *viz.*, Hirakud and the Gandhisagar have very low mean depths of 11.3 which is one tenth of Gandhisagar in area has a mean depth of 32 m.

The mean depth, however, does not show any direct correlation with productivity, either at primary or fish level. Vidur, despite being one of the shallowest (mean depth 2.1 m) of all reservoirs in the country, does not support rich plankton community. Likewise, Kulgarhi and Govindgarh reservoirs in Madhya Pradesh exhibit propensities towards oligotrophy in spite of their shallowness. On the other hand, Gobindsagar, the deepest reservoir, has the highest productivity among large reservoirs. Medium deep reservoirs like Amaravathy (13.7 m), Aliyar (16.8 m), and Tirumoorthy (11 m) develop regular blooms of plankton.

2.2.4. Shoreline and Volume Development Indices

A highly crinkled shoreline, as indicated by the high values of shoreline development index, is believed to be indicative of productive nature of the water body. An irregular shoreline encompasses more littoral formations and areas of land and water interface. High shoreline indices of Hirakud (13.5), Gobindsagar (12.26), Tilaiya (9.12), Konar (8.78), Nagarjunasagar (7.89) and Rihand (7.04) are accompanied by a moderate to rich plankton community.

Ratio between the maximum depth and mean depth, often described as volume development index denotes the depth of basin in relation to the nature of basin wall. An index value less than 1 suggests basin wall convex towards water. No perceptible relation exists between this parameter and the productivity of Indian reservoirs.

Hydrographic changes have a direct bearing on productivity, as sudden changes in water level, inflow and outflow directly affect the biotic communities. It has been observed that plankton, benthos, and periphyton pulses coincide with the months of least level fluctuations and all these communities are at their ebb during the months of maximum level fluctuations and water discharge. Percentage of shallow areas (littoral formation) which varies at different levels, depending on the contour, is also an indicator of productive nature of the lakes.

Storage and release of water from dams are dictated by the requirements of irrigation, power generation and other primary purposes of the dam, rather than any considerations related to fisheries. The spillway discharge, apart from dislodging the standing crop of plankton, removes the oxygenated clear water at the top layer, leaving the oxygen-deficient, turbid bottom water. Similarly, the deep drawdown removes the decomposing material including nutrients.

2.2.5 Hydro-edaphic Factors

The oligotrophic tendencies shown by some of the reservoirs are mainly due to the poor nutrient status and other chemical deficiencies. In most of the cases, poor water quality is a direct reflection of the catchment soil. All reservoirs in Kerala portray a low status in terms of specific conductivity (<50 µmhos) and total alkalinity (<50 mg l^{-1}) with the attendant low primary productivity and plankton.

The rivers of Kerala drain the hills of Western Ghats with lateritic and humus soils deficient in N, P and Ca. The eastern slopes of these hills drain the rivers feeding Hope Lake, Manimuthar, Pechiparai and Peruchani, all deficient in ions. Sathanur, Krishnagiri and Vidur reservoirs receiving drainage passing predominantly through cultivated area have higher levels of alkalinity and hardness. Similarly, in Madhya Pradesh the water is soft to medium soft with less mineral salts, due to geo-chemical reasons. Even small lakes with shallow bottoms, more often than not, do not show signs of productivity due to the poor chemical make up of the catchment. Soils in Madhya Pradesh are normally deep black, medium black, shallow black and mixed red and skeletal, low in nitrogen and phosphorus.

Catchment of Ravishankarsagar comprises rocky, denuded forests and upstream rivers are intercepted by impoundments, which further deprive the water of the suspended matter. Allochthonous enrichment with minerals and nutrients of the reservoir is very low, resulting in low standing crop of plankton. Limestone and other calcareous rocks underlying the water course in the Deccan plateau are responsible for the predominantly hard water character of many of the reservoirs on the Krishna and Cauvery in Andhra Pradesh and Tamil Nadu. The acidic nature of the water of the north eastern reservoirs Kyrdemkulai, Nongmahir and Barapani is attributable to the acidic soil of the reservoir bed and in the catchment.

2.2.6. Nutrient Budget

Most of the Indian reservoirs are characterised by low levels of phosphate and nitrate. Phosphate very seldom exceeds 0.1 mg l^{-1} in reservoirs free from pollution. However, the reservoirs of Rajasthan have particularly high levels of phosphate, ranging from traces to 0.929 mg l^{-1}. They receive phosphate from the rain washings derived from soils types like brown hills, grey brown hills, red and yellow and desert soils. In the highly eutrophic reservoir of Mansarovar in Madhya Pradesh, phosphate levels of 4 to 13 mg l^{-1} were recorded.

Nitrate nitrogen in water in Indian reservoirs is mostly in traces and seldom exceeds 0.5 mg l^{-1}. Lack of nutrients in water, especially the nitrate nitrogen and phosphate, does not seem to be indicative of low productivity. In many cases, despite their virtual absence, the production processes are not hampered. In Amaravathy, Bhavanisagar, Gandhisagar, Ravishankarsagar and many other reservoirs, moderate to very high primary productivity is reported, although the phosphate in water is either absent or present in a very low concentration. In the tropical reservoirs, phosphate level in water has limited scope as an indicator of productive traits. This phenomenon is attributed to rapid turnover of nutrients (Ehrligh, 1960; Abbot, 1967) and their quick recycling, as seen from the high metabolic rates. Hayes and Phillips (1958) showed that 95 per cent of the phosphorus could be taken up by the phytoplankton within 20 minutes, while some algae could convert inorganic phosphate into organic state in less than one minute.

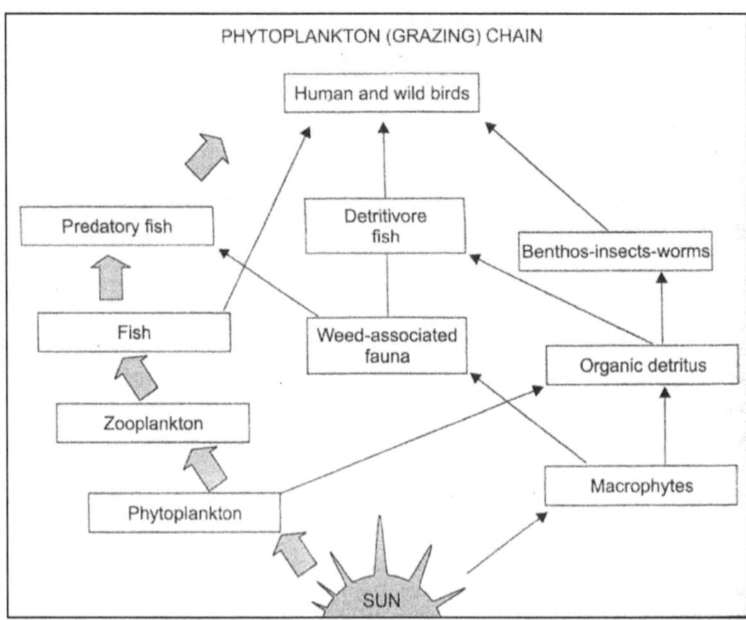

Figure 10: Grazing Chain in Plankton-based Energy Transfer.

Unlike the nutrients like phosphate and nitrate, the measure of total dissolved solids in the form of total alkalinity and the specific conductivity reflects the production propensities of reservoir satisfactorily, with the exception of

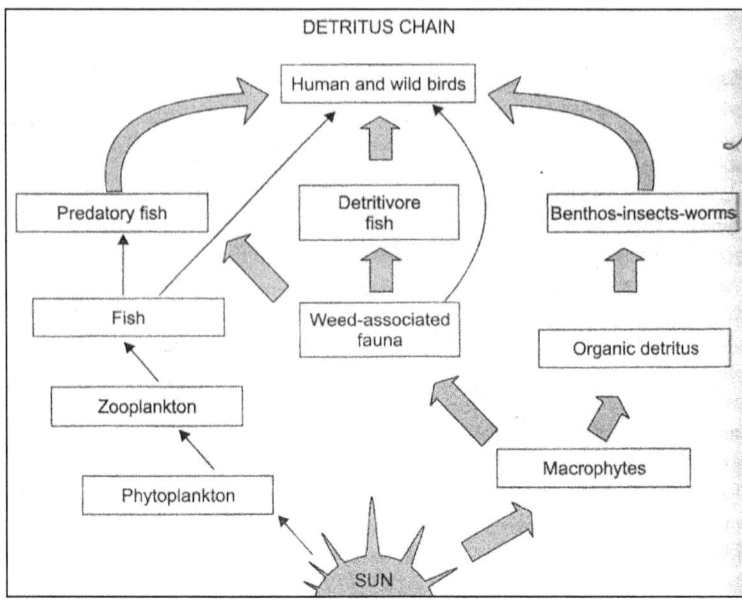

Figure 11: Detritus Chain in Macrophyte-based Energy Transfer.

Amaravathy which, despite very low levels of specific conductivity (38 to 63 µmhos), total alkalinity (7 to 84 mg l^{-1}) and total hardness (18 to 50 mg l^{-1}), supports a very rich plankton community and a good stock of fish.

The ranges of notable abiotic factors indicating their productivity status are given in Table 13. A close examination of physico-chemical data pertaining to more than 100 reservoirs in the country leads to the conclusion that none of the morphometric, edaphic, and water quality parameters can be used as a dependable yardstick to predict the organic productivity to any degree of accuracy, production propensities of each reservoir being determined by a variety of factors. The vertical gradient of some of the chemical parameters, especially the dissolved oxygen, however, conveys the status with a higher level of accuracy.

Table 13: Physico-chemical Features of Indian Reservoirs (range of values)

Parameters	*Overall Range*	*Productivity*		
		Low	*Medium*	*High*
A. WATER				
pH	6.5–9.2	<6.0	6.0–8.5	>8.5
Alkalinity (mg l^{-1})	40–240	<40.0	40–90	>90.0
Nitrates (mg l^{-1})	Tr.-0.93	Negligible	Upto 0.2	0.2–0.5
Phosphates (mg l^{-1})	Tr.-0.36	Negligible	Upto 0.1	0.1–0.2
Specific conductivity (µmhos)	76–474	Upto 200	>200	
Temperature (°C)	12.0–31.0	18	18–22	>22
		(with minimal stratification : *i.e.*,>5°C)		
B. SOIL				
pH	6.0–8.8	<6.5	6.5–7.5	>7.5
Available P (mg/100 g)	0.47–6.2	<3.0	3.0–6.0	>6.0
Available N (mg/100 g)	13.0–65.0	<25.0	25–60	>60.0
Organic carbon (per cent)	0.6–3.2	<0.5	0.5–1.5	1.5–2.5

After Jhingran, 1990.

The bacterial decomposition of organic matter at the bottom is well reflected by the high rate of oxygen consumption. A corresponding increase in oxygen at the trophogenic upper zone gives clues of the high rate of photosynthesis. Almost all productive reservoirs in the country, irrespective of their geographic location, have a klinograde oxygen curve.

In most of the cases, the oxycline is accompanied by a vertical stratification of other chemical parameters such as pH, carbon dioxide, total alkalinity and specific conductivity. The tropholytic zone has a steady supply of free carbon dioxide, which reacts with carbonate to produce bicarbonates. This results in an increase of bicarbonates towards the bottom. Similarly, due to the increase in the hydrogen ions, the pH drops rapidly. Thus, the increase in total alkalinity, specific conductivity and CO_2 and the decrease in pH values towards the bottom layers act as useful indicators of productivity.

Rate of primary productivity in reservoirs is very high due to the warm tropical conditions available in most parts of the country. Many workers consider 1 per cent of the total carbon produced at the phytoplankton phase as the potential fish production from a water body, although almost all reservoirs produce much less fish than their potential.

Chapter 3

Limnological Profile of Indian Reservoirs

Indian reservoirs are situated,by and large at a tropical regime with rich nutrient status conducive for good organic Productivity.Narrow range of fluctuations in water and air temperature during different seasons, a phenomenon which prevents the formation of thermal stratificationMany reservoirs in the Upper Peninsula undergo transient phases of thermal stratification during summer, but wind-induced turbulence churns the reservoirs facilitating the availability of nutrients at the trophogenic zone. Plankton, benthos, and periphyton pulses of Indian reservoirs coincide with the months of least level fluctuations and water discharge. Oligotrophic tendencies shown by some of the reservoirs in the Western Ghats and the Northern are mainly due to poor nutrient status and other chemical deficiencies

3.1. Biotic Communities

The highly seasonal rainfall and heavy discharge of water during the monsoons result in high flushing rate in most of the reservoirs which does not favour colonisation by macrophytic communities. Similarly, inadequate availability of suitable substrata retards the growth of periphyton. Plankton, by virtue of drifting habit and short turnover period, constitutes the major link in the trophic structure and events in the reservoir ecosystem. A rich plankton community with well-marked several succession is the hallmark of Indian reservoirs.

3.2. Plankton

Blue-green algae from the mainstay of plankton community in vast majority of the man-made lakes studied. The overwhelming presence of *Microcystis aeruginosa* in Indian reservoirs is remarkable. The productive water of Gangetic plains, Deccan plateau, south Tamil Nadu and Odisha invariably have good standing crop of *Microcystis*. A common feature of all these reservoirs is the bright sunshine, isothermal water column, klinograde oxygen curve and an extensive catchment area, draining a calcium rich, forested or cultivated land. The species is almost ominipresent in the southern peninsula, except in the reservoirs of Karnataka and Kerala, which tend to be oligotrophic and have poor plankton count with desmids and other green algae as the main constituents. Reservoirs of Rajasthan receiving scanty rainfall and poor flushing rate favour macrophytes and despite being productive do not harbour blooms of *Microcystis*. The oligotrophic lakes of the north-east have a desmid-dominated plankton community (Figure 12).

Altitude plays a decisive role in the distribution of *Microcystis*. Gobindsagar, the highly productive high altitude temperate reservoir, supports a rich community of *Ceratium* sp. instead of *Microcystic*. Similarly, the tropical Markonahalli situated at an altitude of 731 m above MSL, has *Ceratium* sp. as the major constituent of plankton.

Most of the reservoirs have three plankton pulses conciding with the post-monsoon (September to November), winter (December to February) and summer (March to May) seasons. The monsoon (June-August) flushing disturbs and often dislodges the standing crop of plankton. However, no sooner the destabilising effects wean away (as the dam outlets are closed), the allochthonous nutrient input favours an accelerated plankton growth. As the post-monsoon merges into winter, the turbulence decreases and water becomes clean, the plankton community progresses through a series of several successions to culminate in a peak. The summer maxima coincide with the drastic drawdown, bringing the deep, nutrient-rich areas into the fold of tropholytic zone. The temperature, bright sunlight and rapid tropholytic activities also accelerate the multiplication of plankton during summer. In some cases, only two pulses (*i.e.*, the post-monsoon and summer) are seen. However, the shallow, nutrient-rich reservoirs in the southern tip of the peninsula, by virtue of the fast turnover of nutrients and availability of sunshine and warmth, sustain a permanent bloom of plankton.

The ubiquitous blooms of *Microcystis* in reservoirs in peninsular India are an example of a lacustrine biocoenose giving way to fluviatile ones in an impoundment. Studies have indicated that Chlorophyceae and Bacillariophyceae constituted the main components of riverine plankton. On reservoir formation and the consequent transformation of lotic environment into the lentic system, saprophobes disappear from the scene giving room for the rapid multiplication of saproxenes. *Microcystis*, finding a favourable note with the new environment, bursts into blooms, outnumbering all other forms into in significance. In many reservoirs, orientation of lacustrine and fluviatile plankton can be clearly discerned from the composition of plankton in lotic, lentic and the cove sectors. The fluviatile lotic sector, although recording a lower plankton density, often shows better diversity

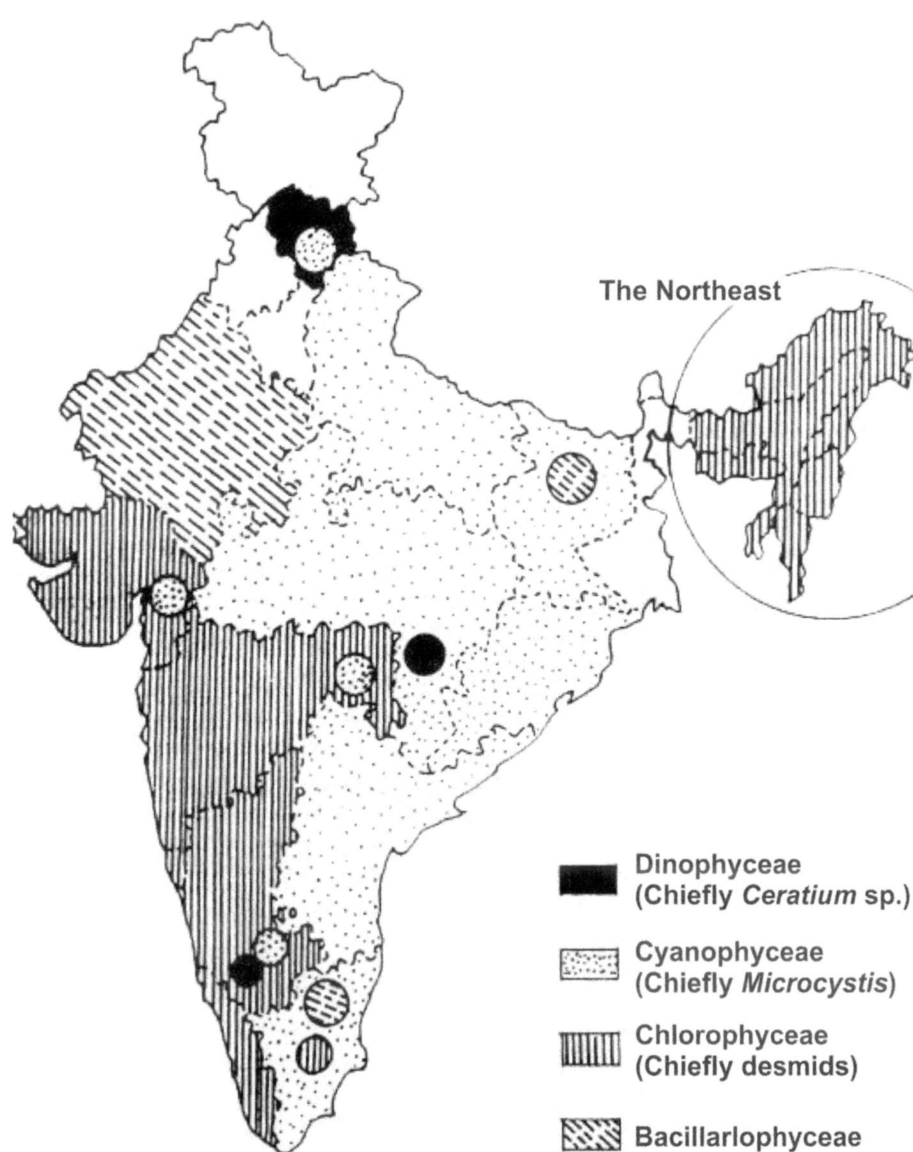

Figure 12: Dominant Phytoplankton in Reservoirs (State-wise).

and evenness indices, compared to the lentic and bay sectors, the still waters of which are characterised by higher concentration of dominance (C) and low evenness(J') (Sugunan, 1991).

3.3. Macrovegetation

Aquatic macrophytes do not figure prominently in the community structure and trophic events of the reservoirs in India, barring some exceptional circumstances, such as low water renewal, ageing of reservoir and pollution stress. In most of the reservoirs they are totally absent or their population is too insignificant to be taken into account. Mostly, they are restricted to isolated patches of *Vallisneria, Hydrilla* and mats of *Spirogyra*, found in the protected bays and coves. Yerrakalava in Andhra Pradesh, Ramgarh in Rajasthan and Sayajisarover in Gujarat are the examples of macrophytic growth due to low flushing rate. Small irrigation reservoirs in Uttar Pradesh. *viz.*, Bachhra and Baghla are also known for the luxuriant growth of macrophytes.

Hussainsagar in Andhra Pradesh and Mansarover in Madhya Pradesh harbour thick vegetation which thrives due to the hyper-eutrophication. These reservoirs are well-advanced on their way to change to swamps. Age of the reservoir seems to have an influence on the macrophyte community. Vanivilas Sagar, and Markonahalli reservoir in Karnataka formed in 1901 and 1939 have a luxuriant growth of macrophytes. Similar age-related macrophyte growth can be observed in Yerrakalava, Ramgarh, Jaisamand, Hussainsagar and Fatehsagar reservoirs. Reservoirs of Rajasthan exhibit a seasonal rhythm in aquatic weeds, their population peaking in summer and declining during the monsoon season.

In the irrigation tanks of Karnataka, the qualitative and quantitative distribution of macrovegetation depends, to a large extent, on the physiographic divisions and the soil types. Thick vegetation comprising the littoral, submerged and emergent types is the characteristic feature of the tanks of the coastal plains and the Malnad region. In the transitional zone between plateau and hills, marked by the presence of laterite and red soil, the tanks are fertile and plankton-rich. Weeds are sparse and whenever present, do not choke the water. In the tanks in the black soil zones, the weeds are mostly submerged and emergent types such as *Hydrilla, Chara Nitella etc.* In most of the tanks in the red soil area, which are seasonal, the vegetation is limited to littoral areas.

Macrophytes offer substrata for an array of insects, molluscs and other invertebrate fauna, and thereby contribute to the species diversity of a water body. Nevertheless, the presence of weeds is considered to be undesirable from fisheries point of view. They accumulate large quantities of inorganic nutrients early in the season, depriving the phytoplankton of their share of nutrients. The floating vegetation utilizes the incident solar radiation for its photosynthesis and makes it unavailable to the phytoplankton communities. Submerged weeds provide shelter for minnows and weed fishes which compete with major carps for food. Excessive growth of macrophytes cause high rate of decomposition of dead plants at the bottom, creating anaerobic conditions. Problems are further confounded, if the water surface is matted by the floating vegetation which prevents light penertration. Instances of fish mortality in summer under such circumstances are reported from Hussainsagar and the reservoirs of Rajasthan. A major deleterious effect of weeds is its physical obstruction they cause to a variety of fishing gear.

The submerged plants commonly encountered in Mansarovar are *Ceratophyllum demersum, Potamogeton crispus, P. perfoliatus, P. pectinatus, P. natans, P. nodosus, Najas graminaea, N. minor, Hydrilla verticillata, Vallisneria spiralis, Chara intermedia, Nitella gracilus.* Floating weeds comprising *Eichhornia crassipes, Lemna sp.* and *Spirodela* and the emergent *Ipomoea aquatica, I. reptans, Polygonum glabrum, Typha angustata*, and *T. elephantia* are reported. Hussainsagar is choked mainly with the water hyacinth, while the reservoirs of Rajasthan harbour a rich flora, dominated mainly by *Hydrilla, Ceratophyllum, Potamogeton, Najas, Azolla* and *Ipomoea*. The marginal areas of Yerrakalava reservoir in Andhra Pradesh have *Hydrilla* and *Chara*, while the deeper zones harbour *Vallisneria* and *Spirogyra.*

3.4. Benthos

Benthic invertebrates fauna show an erratic distribution in Indian reservoirs. The main factors that retard this community are the predominantly rocky bottom, frequent water level fluctuation and the rapid deposition of silt and other suspended particles. In spite of this, a number of reservoirs harbour rich communities of benthic invertebrates. The sequence of dominance of benthic communities closely follows the soil fertility pattern, the pre-impoundment debris often providing suitable habitats.The benthic community succession especially that of chironomids is sometimes used to characterize habitat changes. High shoreline development, variable slopes and vegetation act as favourable factors for the development of a rich assemblage of benthic organisms.

Small, shallow reservoirs of the Gangetic basin, such as Bachhra and Baghla are particularly rich in benthic fauna mainly due to the favourable substratum, rich in organic matter and the absence of swift changes in water level. In Bachhra reservoir, the standing crop of benthos registered a steady growth from 490 to 1894 individuals m^{-2} during the last 10 years. Baghla, another small reservoir in Uttar Pradesh, has a population of benthic invertebrates represented by *Chironomus*, annelids and molluscs. The deep Rihand reservoir in the Ganga basin has a poor benthic community.

Reservoirs of Karnataka, such as Tungabhadra, Markonahalli, Hemavathy, Vanivilas Sagar and Krishnarajasagar have impressive populations of benthic organisms. So are the reservoirs of Himachal Pradesh and Rajasthan. Trends in Tamil Nadu and Madhya Pradesh are erratic. The local conditions, rather than a general geo-climatic features of the area, determine the density of benthic populations in their reservoirs.

Chironomid larvae, being a saprophobic, quickly fill the niches vacated by the saproxenes during the tranformation of habitats. They form the most important constituent of benthos, reported from all soil types and geographic locations and depths. Gastropods and annelids from the next important groups. *Viviparus bengelensis* enjoys country-wide distribution.

Among the biotic communities of the reservoir ecosystem, periphyton is the least reported upon. It constitutes an important component of food for the browsing fishes which contribute substantially to the total fish biomass of the tropical reservoirs. Apart from the limited littoral region in reservoirs, it is the frequent level fluctuations that prevent the growth of periphyton on natural substrata. Significantly, rich periphyton, whenever reported, coincides with relatively stable reservoir levels. There are reports of rich periphyton deposits on anchored boats, rafts, *etc.* that move down along with the receding water level. The fixed substrata either get totally exposed when water level decreases or they are submerged too deep for the communities to survive when level goes up. Propensities for rich settling rates of periphyton have been established through experiments with the artificial substrata, such as glass slides (David *et al.,* 1975; Jha, 1979; Sugunan and Pathak, 1986).

3.5. Ichthyofauna

Despite the cataclysmic faunistic changes associated with the impoundment, Indian reservoirs preserve a rich variety of fish species. The ichthyofauna of a reservoir basically represents the faunal diversity of the parent river system. On the basis of studies conducted so far, large reservoirs, on an average, harbour 60 species of fishes, of which at least 40 contribute to the commercial fisheries. The fastgrowing Indo-Gangetic carps, popularly known as Indian major carps, occupy a prominent place among the commercially important fishes. More recently, number of exotic species have contributed substantially to commercial fisheries. Broad categorisation of the species is as follows:

The Indian Major Carps

Labeo rohita, L. calbasu, L. fimbriatus, Cirrhinus mrigala, Catla catla,

The Mahseers

Tor tor, T. putitora, T. khudree, Acrossocheilus hexagonolepis

The Minor Carps

It is including snow trout and peninsular carps:

Cirrhinus cirrhosa, C. reba, Labeo kontius, L. bata, Puntius sarana, P. dubius, P. carnaticus, P. kolus, P. dobsoni, P. chagunio, Schizothorax plagiostomus, Thynnichthyes sandkhol, Osteobrama vigorsii.

Large Catfishes

Aorichthys aor, A. seenghala, Wallago attu, Pangasius pangasius, Silonia silondia, S. childrenii.

Featherbacks

Notopterus notopterus, N. chitala.

Airbreathing Catfishes

Heteropneustes fossilis, Clartas batrachus,

Murrels

Channa marulius, C. striatus, C. punctatus, C. gachua,

Weed Fishes

Ambassis nama, Esomus danrica, Aspidoparia morar, Amblypharyngodon mala, Puntius sophore, P. ticto, Oxygaster bacaUa, Laubuca laubuca, Barilius barila, B. bola, Osteobrama cotio, Gadusia chapra.

Exotic Fishes

Oreochromis mossambicus, Hypophthalmichthys molitrix, Cyprinus carpio specularis, C. carpio commun, Gambusia affinis, Ctenopharyngodon idella.

Most of the catfishes, featherbacks, air breathing fishes,murrels and the weed fishes enjoy a country-wide distribution, while that of the major carps, minor carps and mahseers (*Tor putitora, T. tor, Acrossocheilus hexagonolepis*) varies according to river basins. The Indian major carps, catla (C. *catla*), rohu (L. *rohita*) and mrigal (C. *mrigala*) constitute the important native ichthyofauna of the rivers of the Gangetic system. These rivers also harbour *Labeo bata, P. sarana, P. chagunio*, and *C. reba, Tor putitora, Labeo dero* and the snow trouts (*Schizothorax* spp.) form the dominant riverine fish fauna of Indus system. Mahseers, especially the chocolate mahseer, *Acrosscheilus hexagonolepis* are also found in the streams associated with all the major river systems of the country. Indigenous fishes of the peninsular rivers include *Cirrhinus cirrhosa, C. reba, Labeo kontius, L. fimbriatus, P. dubius, P. sarana. P. carnaticus, P. kolus, P. dobsoni, T. tor, T. sandkhol* and *O. vigorsii.*

Fish faunistic diversity of a reservoir at a given time is the result of the impact of a series of man-made and natural changes on the native fauna of the parent river. Riverine fish fauna is subjected to a series of habitat changes such as water current, turbidity levels, fishing pressure, loss of breeding grounds and the changes in fish food organisms due to lake formation. The original fauna changes and hardy fish species take advantage of the vacant niches. In many reservoirs, transplantion of fishes from other basins and introduction of exotic species have led to further radical changes in the species set up.

The three Indian major carps have been stocked extensively in reservoirs all over the country for many decades and, in many instances, they have established themselves in reservoirs far away from their original habitat. Sathanur reservoir in Tamil Nadu has a naturalised population of catla that contribute 80 to 90 per cent of the total catch. This Indo-Gangetic carp has eclipsed all indigenous fish fauna including *Labeo fimbriatus*, which dominated the scene by contributing 36 per cent of the catch during the mid 1960s. Similarly, introductions of the silver carp in Gobindsagar, common carp in Krishnarajasagar and tilapia (*Oreochromis mossambicus*) in Amaravathy are examples of man induced changes in fish communities.

3.6. Chemical Stratification

Aside from physico-chemical characters of water and soil as discussed above, the vertical distribution of some of the chemical constituents of water acts as a reliable index of productivity of reservoirs. The upper trophogenic layer of water is characterized by high rate of photosynthesis due to presence of light in contrast to lower tropholytic layer. In upper photosynthetic zone, carbon dioxide is taken up from bicarbonates by photosynthetic organisms and oxygen is liberated during photosynthesis. Thus, trophogenic zone is characterized by increase in oxygen and decrease in carbon dioxide and bicarbonates. In tropholytic zone, oxygen is consumed for purification of organic matter. A strong decline in oxygen (oxycline) and pH coupled with an increase in bicarbonates and carbon-dioxide towards bottom of water indicates high rate of production process in a lake. Oxygen and carbon-dioxide are two complementing agents in metabolism influenced by temperature. When carbon-dioxide in epilimnion decreases, an increases in oxygen is expected. On the contrary to this process, when carbon-dioxide in hypolimnion increases, there is decrease in oxygen level (Klinograde curve). Thus, productivity of a lake is estimated from the nature of the oxygen curve. The klinograde curve of oxygen is considered to be productive character against orthograde when oxygen value is more or less uniform from surface to bottom.

3.7. Trophic Dynamics and Energy Flow

A reservoir is known to pass through three distinct phases after its formation, *viz.* initial high fertility phase, trophic depression phase and final fertility phase. A newly formed reservoir inundates vast areas of forest and agricultural lands causing decay of submerged vegetation. This results in release of nutrients causing initial fertility leading to intense development of fish food organisms - plankton, bottom microflora and fauna. This stage lasts for 2-3 years and is followed by trophic depression stage, caused by rapid utilization of nutrients by flora and subdued release of nutrients from reservoir-bed due to sedimentation. This phase is marked by low production of fish –food organisms and lower fish growth, hence lesser production. In Indian reservoirs, trophic depression generally exists for shorter duration and it varies from reservoir to reservoir. After depression period, the reservoir gradually recovers with accumulation of nutrients. The final fertility of reservoir on stabilization is somewhere near half the magnitude of initial phase,which gets adjusted to the basic productivity of basin depending on the watershed runoff and inflow. But in Indian reservoirs it has been shown that final fertility would be of much higher magnitude than initial and reservoir productivity improves after ageing.

A basic knowledge of aquatic ecology is helpful in managing production aspects of reservoirs. The aquatic ecosystem consists of biotic communities of producers (autotrophs), consumers (heterotroophs) and decomposers (saprophytes). The phytoplankton is the largest group, involved in autotrophic production processes, called primary producers, being in the first trophic level and controlling basic productivity or 'primary production' of the ecosystem. The primary consumers are in the second trophic level represented by zooplankton and some fishes (grazers and browsers). Secondary consumers or carnivores are in the third trophic level, including insect larvae, fishes *etc.* Tertiary consumers belong to next higher trophic level represented by large predators or top carnivores. The decomposers, bacteria, fungi, also play a leading role in mineralizing organic matter and recycling nutrients. During energy transformation operating on the above trophic pyramid, the energy fixed by primary producers passes through different trophic strata and efficiency of the conversion is reduced to a tenth from one level to the next higher level.

3.8. Energy Flow

The biotic communities (producers, consumers and decomposers) in an ecosystem are linked with one another with energy chains. Energy flow and nutrient cycles are two important principles of ecology. Complete knowledge of inter relationships among organisms, energy flow and nutrients from one level to other and role of environmental parameters in energy transformation processes are very important for understanding ecosystem. This study of productivity is now receiving much importance and ecologies are keen to know the efficiency with which solar energy is converted to chemical energy is utilized by consumers (ecological efficiency), and efficiency conducted in Bhavanisagar, Nagarjunasagar, Rihand and Gobindsagar reservoirs have shown that photosynthetic efficiency was high in productive reservoirs like Bhavanisagar (0.412 per cent), Nagarjunasagar (0.290 per cent) and Gobindsagar (0.682 per cent) but low in unproductive Rihand (0.202 per cent). It is interesting to note that energy harvest as fish was much low in Nagarjunasagar (0.055 per cent) as compared to Bhavanisagar (0.290 per cent), which solved that management failed in harvesting fish from Nagarjunasagar though it was productive.

Productivity of a lake is dependent on biogenic capacity to transform solar energy into chemical energy. The energy fixed at primary level passes through trophic chain and fraction of it ends up as fish flesh (trophic dynamics). Therefore, the structure of different food biotic communities assumes great significance to reservoir fishery management. Community metabolism or the transfer of energy from one trophic level to the other can be a major criterion for selecting management options, especially the species selection in culture-based fisheries. In an ecosystem, the biological output or the production of harvestable organisms can be at various trophic levels. Under a grazing chain, a phytoplankton > zooplankton > minnows > catfishes systems or a phytoplankton > zooplankton > fish system prevails. In macrophyte dominated system, the primary energy at the macrophytes level is invariably channeled through detritus chain. There are different detritus chains such as macrophytes >detritus> detritivore system, phytoplankton > detritus > benthos > bottom feeders system and macrophytes associated fauna > air-breathing fish system. Shortning food-chain will lead to higher rates of fish production but in reservoir management there is little scope for changing community structure of plankton to increase primary productivity. However, alternations in species spectrum of fish may be done, and for this only stocking of fish is a successful tool in management

Chapter 4

Management of Small, Medium and Large Reservoirs

4.1. Reservoir Fisheries

A realistic evaluation of fish production from reservoirs in India is elusive. Compared to the impressive volume of data generated by the individual research workers and various institutions on limno-chemical variables and biotic communities, the estimates on fish catch remain grossly inadequate. Reliable fish catch statistics and yield estimates remain as the weakest link in the database on reservoirs. More discomfitting is the fact that the production figures available on most of the reservoirs are inaccurate and unreliable. This lacuna is imputable to a large number of factors, chiefly:

1. The multiplicity of agencies owning the fishing rights that pose difficulties in some States to gather data.
2. Highly scattered and unorganised market channels, mostly under the clutches of illegal money lenders.
3. Ineffective cooperative set up,
4. Diverse licensing/royalty/crop sharing systems practised by different State Governments, some of which include a *free for all* system, providing little scope for recording catch statistics, and
5. Inadequate and poorly trained manpower at the disposal of State Governments/Cooperatives to collect catch data, follow statistically sound sampling procedures, unable to cover the whole reservoir.

All India Coordinated Project on Reservoir Fisheries took the lead in evolving a collection methodology on the basis of stratified random sampling during 1971 to 1985. However, many State Governments were not able to follow the procedure and to continue with recording of catch data after the project wound up in 1985. Nevertheless, some States like Tamil Nadu and Madhya Pradesh, despite the enormity of the resource size, have a streamlined machinery to record catch statistics. Similarly, Himachal Pradesh has good documentation on catch. On the contrary, Karnataka, has very little information on the catch structure of its reservoirs. Figures of Andhra Pradesh seem to be off the mark, due to the free fishing system and inadequate methods of data collection. For instance, system followed in Nagarjunasagar virtually allows anybody to catch fish in the reservoir and freely sell it anywhere. Considering the size of the lake and the large number of remote landing centres, an effective monitoring of catch is almost impossible. This is equally true with regard to many reservoirs of Kerala, Maharashtra, Odisha, Karnataka and Uttar Pradesh.

Based mainly on the data obtained from various State Governments, the fish production particulars from 422 reservoir have been presented in Table 14. Fish yield figures of small reservoirs of Andhra Pradesh, as given by the State Fisheries Department are very impressive (188 kg ha^{-1}), followed by those of Kerala, Madhya Pradesh, Tamil Nadu and Rajasthan in the range of 46.43 to 53.5 kg ha $^{-1}$. Medium reservoirs of Rajasthan, on an average, produce fish at the rate of 24.47 kg ha^{-1}, while Tamil Nadu, Maharashtra, Madhya Pradesh and Odisha record about half this yield. The two large reservoirs in Himachal Pradesh produce 35.55 kg ha^{-1} which is remarkably high, compared to other States.

The estimated fish yield of 291 small, 110 medium and 21 large reservoirs of the country are 49.90, 12.30 and 11.43 kg ha^{-1} respectively. Based on the catch of 422 reservoirs belonging to 10 states during 1992–93, the national fish production rate of Indian reservoirs is estimated as 20.13 kg ha^{-1}. Applying this national average yield rates into the 1 485 557 ha of small, 527 541 ha of medium and 1 140 268 ha of large reservoirs in the country, their current production rate can be estimated as 74 129, 6 488 and 13,033 trespectively. A modest increase in yield rate up to 100, 75 and 50 kg ha^{-1} in respect of small, medium and large reservoirs, would ensure production of 148 556, 39 565 and 57,013 t. This would increase the production by 2.5 times *i.e.*, from the present 93 650 t to 245 134 t (Table 15).

4.2. Reservoir Fisheries Management in India

The present low level of fish production in Indian reservoirs can be attributed to inadequate management as much as many of them have high propensities of production from a limno-chemical point of view. In many of the reservoirs, the high rate of the primary and secondary productivity is not channelled to fish production. Insufficient understanding of the reservoir ecosystem often comes in the way for adopting effective management measures.

Table 14: Fish Production in different Categories of Reservoirs in India

State	Small Reservoirs			Medium Reservoirs			Large Reservoirs			Pooled		
	No.	Production (t)	Yield (kg ha^{-1})	No.	Production (t)	Yield (kg ha^{-1})	No.	Production (t)	Yield (kg ha^{-1})	No.	Production (t)	Yield (kg ha^{-1})
Tamil Nadu	52	760	48.50	8	269.0	13.74	2	294.0	12.66	62	1323.0	22.63
Uttar Pradesh	31	168	14.60	13	156.0	7.17	1	50.0	1.07	45	374.0	4.68
Andhra Pradesh	37	2224	188.00	29	306.0	22.00	3	800.0	16.80	69	4330.0	36.48
Maharashtra	6	72	21.09	12	313.5	11.83	4	794.0	9.28	22	1179.6	10.21
Rajasthan	78	970	46.43	17	599.7	24.47	2	120.0	5.30	97	1690.0	24.89
Kerala	7	118	53.50	2	17.3	4.80	–	–	–	9	135.0	23.37
Bihar	25	22	3.91	3	7.2	1.90	1	0.8	0.11	28	30.0	0.054
Madhya Pradesh	2	24	47.26	20	624.9	12.02	3	1184.0	14.53	25	1833.1	13.68
Himachal Pradesh	–	–	–	–	–	–	2	1453.0	35.55	2	1453.0	35.55
Odisha	53	349	25.85	6	163.0	12.76	3	925.0	7.62	62	1437.0	9.72
Total	291			110			21			422		
Average		49.90			12.30			11.43			20.13	

Table 15: Present and Potential Production from Reservoirs of India

Category	Yield (kg ha^{-1})	Area (ha)	Present Production	Potential Production
Small	49.90	1,485,557	74,129	148,556
Medium	12.30	527,541	6,488	39,565
Large	11.43	1,140,268	13,033	57,013
Total	**3,153,366**	**93,650**	**245,134**	

Since fish production from reservoirs is essentially extractive in nature, the essence of management strategy lies in exploitation of natural stocks. Nevertheless, the ecosystem management provides different degrees of freedom for stock maniputation, depending on the size and class of the water body. One of the possible criteria that can be used to differentiate between capture and culture fisheries is the extent of human intervention in the ecosystem management. While aquaculture systems provide maximum avenues for the man to monitor and change the habitat variables and the biotic communities at will, this freedom attenuates as we proceed from aquaculture to the culture-based and capture fisheries. In a large water body, managed on capture fishery norms, there is little room for altering the habitat variables and the scope for effecting change in biotic communities is limited to stocking and ranching, which have uncertain chances of success.

Relative contribution of culture and capture norms in management vary, depending on the category of the reservoir. Medium and large reservoirs are predominantly capture fisheries systems and the management norms are based on the principle of stock manipulation, adjustment in fishing effort, observance of conservation measures and gear selectivity. Selective stocking is resorted to for correcting imbalances in species spectrum and to fill the vacant ecological niches. The small reservoirs, on the other hand, are generally managed as culture-based capture fisheries, akin to extensive aquaculture, where the main accent is on stocking, fattening and harvesting. An imaginative stocking and harvesting schedule and right species mix hold the key for effective management of small reservoirs.

4.3. Reservoir Fisheries Management

Fisheries management of reservoirs is based on the principles of 'enhancement', which defines as a range of management practices/processes by which qualitative and quantitative improvement is achieved from water-bodies through exercising specific management options. This is something intermediate beween culture and capture fisheries. Enhancement inter alia includes 'culture-based fisheries (stock and recapture)', 'stock enhancement (enhanced capture fisheries)', 'species enhancement (introduction of species)', 'environmental enhancement (fertilizing water-bodies)', 'management enhancement (introducing new management options)' and 'enhancement through new culture systems (cage culture, pen culture, FADs, *etc.*)'. Reservoirs offer scope for one or more forms of enhancement. The most suitable management strategy for a particular reservoir is chosen, based on its morphometric, edaphic and biological characteristics. The two most common forms of enhancement followed in Indian reservoirs are culture-based fisheries and stock enhancement.

Culture-based fisheries is generally practiced in the small reservoirs, while the medium and large reservoirs are managed on the basis of stock enhancement, also called enhanced capture fisheries. However, it must be borne in mind that the area alone is not the criterion for determining whether a reservoir is managed on the basis of culture-based fisheries or stock enhancement. A number of other considerations such as depth, predator pressure and fishability also come into play. A set of broad guidelines for distinguishing small reservoirs, which are suitable for culture-based fisheries, from the medium and large reservoirs that are primarily used for stock enhancement. In addition to their physical attributes, impoundments are defined by factors including species diversity, fertility and structural complexity of the aquatic habitat. The goal of fishery management is to control these factors to produce a harvestable surplus while maintaining a dynamic equilibrium within the ecosystem.

4.4. Pre and Post Impoundment Changes in Improving Reservoir Fisheries

Impounding of any river results into transformation of hitherto fluviatile ecosystem to more or less lentic system and in the process of such modification, comprehensive morpho-ecological changes take place and the adaptability of a host of biotic constituents of the ecosystem succumb to the pressure exerted by the emerged scenario; while a number of these may find congenial environment too. As such, efficient tackling by undertaking the Environmental Impact Assessment (ETA) and framing the judicious management practices becomes the key word towards holistic approach for sustained development and possible conservation of the system under developmental constraints

Impoundment Consequences

Damming of any river manifests in following ways:

A. The dam acts as physical barrier which is detrimental to the migration of fishes restricting their accessibility to breeding, nursery and feeding grounds.

B. Profound morpho-ecological changes both above, and below the dam site take place which include :

   1. Hydrodynamic and hydrographic variations associated with water level fluctuations.

   2. Creation of a large mass of water undergoing three defined trophic phase *viz.*, trophic burst, trophic depression and trophic stablization, which offers avenues for fisheries development.

3. Changes in physico-chemical regime.
4. Flooding of spawning and feeding grounds. At downstream, reduction in flood plains, even the spawning grounds dry up owing to confined freshwater availability or shallow zones are created which are unaccessable.

C Dams commissioned over estuaries or at proximity exert negative impact at downstream due to limited freshwater discharge leading to hypersaline conditions. The anadromous fishery gets a jolt due to changed salt-water balance. Owing to decline in diluting capacity, flushing and transporting of wastes during practically low or no discharge period, the downstream environment will experience severe stress.

4.5. Culture-based Fisheries in Small Reservoirs

Management of culture-based fisheries involves stocking of fish into the reservoir, allowing the stock to grow utilizing the natural fish food resources and harvesting them at an appropriate size. Therefore, the number of fish to be stocked, the size at which they are stocked, the size at which they are stocked, the period of growth and the size at which they are harvested play a key role in the success of culture-based fisheries in small reservoirs. In this regard, the key management decisions to be made are:

1. Estimation of fish yield potential,
2. Selection of fish species for stocking,
3. Stocking rate and size, and
4. Period of growth and size at harvesting. Estimation of fish yield potential

It is essential to assess the fish yield potential of the reservoir for formulating appropriate stocking strategies, especially stocking density. Fish yield potential of the water body is determined by its biotic and abiotic characteristics. Several methods are in vogue to assess the fishery potential of small reservoirs by deriving equations based on area, depth, catchments area and the chemical parameters of soil and water. Among them, the morpho-edaphic index (MEI) method is widely accepted. This method combines the morphometric as well as chemical composition and depth are important parameters under Indian conditions. This method involves calculation of MEI as:

$$\text{MEI} = \frac{\text{Specific conductivity } (\mu\text{mhos/cm})}{\text{Mean depth (m)}}$$

Fish yield potential is then calculated using the formula:

Fish yield in kg/ha/year (C) = 0.9898 MEI 1.3888 Selection of fish species and stocking rate

The selection of species for stocking is guided by chances of the stocked species to thrive in the reservoir and effectively utilize the food resources and converting them into fish flesh at the quickest possible time.

Basic principles that govern selection of species for stocking are:

1. The stocked species should find the environmental suitable for maintenance and growth.
2. It should be a quick growing with high efficiency of food utilization (shortest food chain)
3. The size of the stock should be chosen with the expectation of getting the desired results.

One of the important criteria for stocking policy is to know the amount of food available per individual in the environment. This factor has a considerable bearing on population density hence production. In multi-species systems, fish can occupy different niches where competition is avoided or at least minimized. Species competition for reservoirs under consideration all the three Indian major carps (*Labeo rohita, Catla catla* and *Cirrhinus mrigala*) and the freshwater prawn (*Macrobrachium rosenbergii*) can be ideal candidate species.

Stocking rates need to be fixed for individual water-bodies or a group of them sharing common characteristics such as size, presence of natural fish population, predation pressure, fishing effort, possible stock loss, minimum marketable size and multiplicity of water use. The number of fish to be stocked is based on the growth of individual fish and the total possible yield of the reservoir. In other words, it is a individual fish and the total possible yield of the reservoir. In other words, it is a function of the total biomass of fish and the weight of the individual fish at harvest.

The main consideration in determining the stocking rate is growth of individual species stocked, the mortality rate, size at stocking and the growing time. A formula to calculate the stocking rate (Welcome, 1976) is given below.

$$S = (qP/W)\ e^{-z\ (te-to)}$$

Where S number of fish to be stocked (in number/ha); P, natural annual potential yield of the water body; q, proportion of the yield that can come from the species in question: W, mean weight at capture: te, age at capture; tc, age at stocking: -z, total mortality rate.

P can be estimated through MEI method and the range of mortality rates can be found out from the estimated survival rate. The methodologies for calculating stocking rate as described above are only indicative. Higher stocking densities might be possible due to increase nutrient status of the reservoir through extraneous inputs (*e.g.* higher organic loading), larger catchment area feeding the water body and other favourable conditions (*e.g.* absence of predatory species, *etc*). in such cases, yield rates much higher than what is calculated through the formula is possible.

Stocking Size

The size of stocking is important from economic as well as biological points view. Biologically, the larger the fish stocked the better its chances for survival and thus even lesser number of individuals can be stocked. But growing fish to fingerling size is an expensive proposition and it is economically expedient to stock at a lower size in larger numbers. This can be done only when the ecosystem is free from or has lesser number of predators. The discretion of the manager is therefore important to determine the size at stocking. In any case, fingerlings of >100mm size are always considered to be safe and give the best results. In data-deficient situations, stocking of 80-100mm size fingerlings is always recommended.

Period of Growth and Size at Harvesting

Normally, the stocks are replenished annually in small reservoirs in small reservoirs and thus the growth cycle is annual and the stocked fish grow to 0.7-1.0 kg at harvest depending upon the water quality and species stocked. However, fish grow to a harvestable size (starting from 0.5 kg onwards) in a culture-based fishery in 6-8 months. Thus, the seasonal reservoirs that retain water up 6-8 months can be used for culture-based fisheries. However, it must be borne in mind that the fish harvested at a lower size will give higher yield rate in terms of kilograms/ha. Therefore, depending upon the requirements and acceptance of the market, the size at harvest can be determined.

Staggered Stocking and Harvesting

The practice of staggered stocking and harvesting is known to yield better efficiency and economic returns. It allows replenishment of harvested stock at regular intervals and thereby optimum utilization of the inherent productivity of the water-body. Further, it is also permits catching of fish only inherent above certain size and voluntary release of smaller fishes back into the reservoir. Higher level of motivation and awareness among fisheries is required for practicing this method.

In several small reservoirs of Tamil Nadu, staggered stocking and harvesting is yielding better results and is, therefore, recommended for adoption. However, caution should be exercised to avoid overstocking and subsequent low growth rate due to reduced availability of food.

Stock Loss

Loss of juvenile and adult fish through the overflowing spillways poses a serious problem in reservoirs. The situation is further worsened by heavy escape of fingerlings and adults through irrigation canals. Development of fisheries in such water bodies, therefore, requires suitable screening of the spillway and the canal mouth. Such protective measures have already been installed in some of the reservoirs. However, caution is to be exercised to see that the screens across spillway do not get clogged during flood season, which may threaten to damage the dam. In some of the reservoirs fishes have also been observed to move up the spillway into the reservoir, whereas in others the spillways provide an insurmountable barrier to fish moving up the dam. To minimize losses by way of escape of fish through spillway and canal, it would be an economic proposition to have an annual cropping policy so that the reservoir is stocked in September-October and harvested by June end. However, this depends on the growth of fish and the general productivity of the water body.

4.6. Impact of Culture-Based Fisheries in Small Reservoirs

Scientific studies have shown that fish yield is significantly correlated to stocking. Success in management of culture-based fisheries in small reservoirs depends more on recapturing the stocked fish rather than on their building-up a population. The smaller water-bodies have the advantage of easy stock monitoring and manipulation. Thus, the smaller the reservoir the better the chances of success in the stock and recapture process. In fact, an imaginative stocking and harvesting schedule is the main theme of fisheries management in small, shallow reservoirs. The basic tenets of such a system involve.

1. Selection of the right species, depending on the fish food resources available in the system.
2. Determination of stocking density on the basis of production potential, growth and mortality rates.
3. Proper stocking and harvesting schedules, including staggered stocking and harvesting, allowing maximum grow-out period, taking into account the critical water levels.
4. In case of small irrigation reservoirs with open sluices the season of overflow and the possibilities of water level falling too low of completely drying up, are also to be taken into consideration.

Aliyar reservoir in Tamil Nadu, where culture-based fisheries was tried by CIFRI, is a standing testimony to the efficacy of the staggered stocking. The salient features of the management options adopted in Aliyar were:

1. Stocking limited to Indian major carps (earlier, all indigenous, slow-growing carps were stocked),
2. Increasing the size of stocking to 100 mm and above,
3. Reducing the stocking density to 235-300/ha (earlier rates were erratic ranging between 500 and 2,500/ha).
4. Staggering the stocking, and regulating mesh size strictly and banning catch of Indian major carps < 1 Kg in size.

A direct result of the above management practice was an increase in fish production from 1.67kg/ha in 1965-66 to 194 kg/ha in 1990. Successful stocking has also been reported from a number of small reservoirs in India. In Markonahalli, Karnataka, on account of stocking the percentage of major carps has increased to

61 per cent and the yield increased to 63 kg/ha. Yields in Meenkara and Chulliar reservoirs in Kerala have increased from 9.96 to 107.7 kg/ha and 32.3 to 275.4 kg/ha, respectively, through sustained stocking. In Uttar Pradesh – Bachara, Baghla and Gulariya reservoirs, registered steep increase in yield through improved management with the main accept on stocking. An important consideration in Gulariya reservoir was to allow maximum grow-out period between the date of stocking and the final harvesting, *i.e.* before the levels go below the critical mark. The possible loss due to the low size at harvest was made good by the number. Bundh Beratha reservoir in Rajasthan, stocked with 100,000 fingerlings a year resulted in a fish yield of 94kg/ha; 80 per cent of which constituting catla, rohu and mrigal

4.7. Other Forms of Enhancement in Small Reservoirs

Although culture-based fisheries is considered as the most common form of fisheries enhancement, scope exists for other forms of enhancement such as 'species enhancement.' 'environmental enhancement' 'enhancement through new production systems' and 'integrated production systems'.

4.7.1. Species Enhancement and Exotics

Decline of indigenous fish stocks due to habitat loss, especially that caused by dam construction is a universal phenomenon. All the major river basins have been affected, but the extent of such fish species loss is not assessed to any reliable degree. Planning of economically important, fast-growing fish from outside with a view to colonizing all the diverse niches of the biotope for harvesting maximum sustainable crop from them is species enhancement. The new species so induced into the system could be exotic or indigenous. In case the fish is induced to a place outside its normal range of distribution, the species is called 'exotic' and the act is called 'introduction'. Species enhancement in small reservoirs can be done either through indigenous species or introduction of exotics; the latter is subject to the prevailing rules and regulations of the government.

4.7.2. Environmental Enhancement

Improving the nutrient status of water by selective input of fertilizers is a common management tool adopted in intensive aquaculture. However, a careful consideration of the possible impact on the environment is needed before this option is resorted to in reservoirs. It is generally believed that most of the lakes and reservoirs may have sufficient nutrient inputs any excessive nutrient loading can lead to pollution. However, scientific knowledge to guide the safe application of this type of enhancement and the methods to reverse the environmental impacts, if any, is still inadequate. On account of these, this management tool is not commonly applied in India. China is known to have used this practice in a big way to augment production from small reservoirs. Cuba, taking a cue from china has tried manuring of small reservoirs using both organic and inorganic fertilizers. Thailand has also adopted this practice in a selective manner.

Fertilizers are less effective in soft water with total alkalinity <20 mg/litre. Soft waters have inadequate carbon (usually in the form of carbon-dioxide and bicarbonate) for good phytoplankton production. Hence productivity can often be enhanced by applying lime to low alkalinity-impounded waters. The application of lime equivalent to 2,000 to 6,000 kg/ha calcium carbonate is generally sufficient to maintain total alkalinity above 20mg/litre. Fertilization of reservoirs as a means to increase water productivity by abetting plankton growth has not received much attention in India. Multiple use of the water-body and the resultant conflict of interests among the various water users are the main factors that prevent the use of this management option. Surprisingly, fertilization has not been resorted to even in reservoirs, which are not used for drinking water and other purposes. Documentation on fertilization of reservoirs in India's is scarce. Attempts to improve the plankton productivity of Vidur reservoir by the application of super phosphate gave highly encouraging results. As soon as the canal sluice was closed, 500kg super phosphate with P_2O_5 content of 16 to 20 per cent was applied in the reservoir when the water spread was 50 ha with a mean depth of 1.67m. As an immediate result of fertilization, the phosphate content of water increased from nil to 1.8mg/litre and that of soil from 0.242 to 0.328 per cent. Similar improvements in organic carbon and Kjeldal nitrogen have been reported from the soil and water phases on account of fertilization. Experiments were also conducted with urea in the same reservoir.

Application of lime was tried in some upland natural lakes for amelioration of excessive CO_2 and acidity at the bottom. This measure, together with the application of superphosphate in Yercaud Lake in Tamil Nadu, raised the pH of water from 6.2 to 7.3 and decreased the CO_2 in bottom water from 38 to 6.5 mg/litre. There was a corresponding increase in species number and biomass of plankton. Fertilization in Vidur reservoir resulted in a marked increase in benthic and plankton communities and doubling of the primary production rate. After two successive applications of fertilizers, significant limnological changes took place including the presence of free carbon-dioxide and decrease in pH and dissolved oxygen at the bottom layer of water. The methyl orange alkalinity increased from 44 to 108 mg/litre from the surface to bottom, indicating high organic productivity. Phosphates fertilization triggered the tropholytic activities mineralizing the organic matter and producing carbon-dioxide. As a direct benefit from the fertilization, a 50 per cent increase in fish production, along with three-fold increase in the size (average weight of catla, and rohu, mrigal, *Labeo fimbriatus* and *L. calbasu*) were achieved.

Artificial eutrophication as a decisive management option was tried in India for the first time in Kyrdemkulai (80ha) and Nongmahir (70 ha) reservoirs of the north-east by applying poultry manure (10 tonnes/ha), urea (40ka/ha) and single superphosphate (20kg/ha). Fertilization can play a key role in many small reservoirs of India, which require correction of oligotrophic tendencies. A number of reservoirs in Madhya Pradesh, the north-east and the Western Ghats receiving drainage from poor reassuring. In Shishantou reservoir, a management strategy comprising fertilization by organic and production hike from 1,500 kg/ha to 6,000 to 7,000 kg/ha during 1985 to 1989. Before fertilization, the plankton biomass in Shishantou was 1.5 mg/litre, which was raised to 6.5 mg/litre through application of organic fertilizers at the rate of 6,375 tonnes/ha. The plankton biomass, after dropping during the peak precipitation period, picked up to 20.51 mg/litre during the post-rainy season months, with corresponding increase in fish production.

4.7.3. Integrated Production Systems

Small reservoirs are also amenable to integrated aquaculture since the culture-based fishery can be effectively combined with piggery, duckery and poultry rearing. Many of the waste products form these animal husbandry practices act as a fish food fertilizer enabling higher stocking densities and fish yield. However, this approach has limitations from the aesthetic and hygienic point view, especially when the reservoir is a source of drinking water supply. Sometimes, the littoral areas of reservoirs are used for agriculture, especially for farming of soil. Such increase in fish production and rural earnings can make a significant contribution to the nutritional requirements of the rural community.

4.7.4. Stock Enhancement in Medium and Large Reservoirs

Fish production system in majority of medium and large reservoirs is based on the principle of enhanced capture fisheries (stock enhancement) as the essence of management strategy here lies in capture of self-sustaining stocks. This process involves heavy initial stocking, followed by conservation of habitat to allow natural breeding and recruitment, regulated fishing and supportive/corrective stocking whenever necessary. In such water-bodies, the main emphasis would be to build a breeding population that can support a sustainable catch on a regular basis without having to stock annually. At the same time, there is scope for stock manipulation through adjustment in fishing effort, observance of conservation measures and gear selectivity. Selective restocking can also be resorted for correcting imbalances in species spectrum and to fill the vacant ecological niches. It must be borne in mind that the area alone is not the criterion for determining whether stock enhancement is to practiced. A number of other considerations such as depth, predator pressure breeding grounds, habitat and fishability also come into play.

Under an enhanced capture fishery regime, fish productivity depends on a number of factors such as a conductive environment for primary production, a good food web (well balanced biotic communities structure that ensure good conversion of primary energy into fish flesh), self-sustaining fish stock(s) that contributes to high yield and good post-harvest and marketing arrangements. In other words, there is a need for an enabling habitat, good fish stock(s) and good post-harvest arrangements.

Various management processes involved in reservoir fisheries are:

1. Environment management
2. Stock management

3. Fish stock monitoring
4. Craft, gear and effort management
5. Unconventional production systems
6. Other management measures
7. Harvesting and post-harvest management

4.7.5. Environmental Management

The main focus of environment management centres on:

1. A conducive habitat for primary producer organisms to flourish and produce carbon at a high rate, and
2. Maintain the habitat at optional levels to suit the life- cycles of all organisms including fishes. In open waters, the harvestable biotic communities often belong to the nekton, which, in turn, is dependent on plankton, macrophytes or benthos. Any sensitive environmental parameters that adversely affect any of the component communities in the trophic chain are bound to affect the fish production. Thus, the capture fisheries management in reservoirs needs to be environmental friendly, aiming at conservation of the whole food chain.

Eco-degradation of reservoirs has been on the increase due to the rapid pace of industrialization, poor countries management in the catchment and a variety of other factors. Apart from the direct entry of industrial, municipal and thermal wastes, the pollution load carried by the upstream rivers is also accumulated in the reservoirs. The environmental degradation in reservoirs is caused mainly by the waste discharge from industrial, municipal and agricultural sources and the thermal power plants. High rate of siltation due to poor catchment management also affects the biological productivity. In many cases, the water quality of reservoir is influenced by the catchment area situated hundreds of kilometers away from the lake. For instance, soil erosion in the distant catchment area lead to high sediment load and siltation in reservoirs, posing hazards to fish populations. The reservoir fisheries management therefore, entails protection of both terrestrial aquatic ecosystems from environmental degradation.

The major habitat constrains that come in the way of effective reservoir management are siltation and pollution from thermal, domestic, agricultural and industrial sources. Excessive siltation leading to drastic decrease in the water-holding capacity and even damage to concrete hydraulic structures is a common problem in reservoirs. Apart from diminishing the water-holding capacity of the reservoir and shortening its life, siltation also affects the biota by the blanketing the benthic and periphytic community. It also happens the recruitment by destroying the breeding grounds and retards the overall productivity of the ecosystem. A number of reservoirs have been selected, of late, as sites for thermal power plants due to their dual utility as perennial source of water supply and disposal point for heated effluents and fly ash. The main ecological consequences of a hot-water discharged into the aquatic ecosystem are increase in water temperature, change in chemical composition and change in metabolism and life history of aquatic communities. Thick mat of fly ash deposited at the bottom bed over the years seal the nutrients away from the water phase and destroy breeding grounds of fish.

A number of reservoirs contiguous to towns and cities face threat from sewage pollution. Although from the fisheries point of vew, organic loading within certain limits does not hamper the productivity, sewage load in excess can cause aseptic conditions, adversely affect the biotic communities, retard productivity and render the fish unit for human consumption. Moreover, the problem needs to be addressed from public health and aesthetic points of view. Wastes emanating from industries such as chemical plants, textile mills, heavy engineering plants, paper mills, iron and steel factories and rayon are often dumped into the reservoir causing hazards. Hazardous and toxic substances such as pesticides and heavy metals are also catchments. These substance are highly persistent and thereby contaminate the entire bio-geochemical cycle of static like reservoir.

4.7.6. Stock Management

Management of fish stocks in reservoirs entails maintenance of enough quality stocks in adequate numbers to sustain a fishery. This involves several steps such as: stocking, conservation the fish habitat including breeding, dwelling and feeding grounds; fishery gear and effort regulations; closed season; regulations on exotic fishes.

Stocking

Stocking is sine qua non for the reservoir fisheries for building-up stocks of fast-growing species in the ecosystem to colonize the diverse niches. It is important to stock the species that may breed and ultimately get naturalized in the system through autostocking. This is imperative to meet the long-term objective of obtaining a sustained yield. Management involving persistent stocking not only pushes up the input cost, such systems may also create many practical difficulties in raising the stocking material in adequate quantities.

The initial period of reservoir formation (trophic burst) is the right time to stock the desirable species into the reservoir. Heavy stocking with fast growing fishes on a short food chain is essential during this phase along with protection of breeding grounds. This will facilitate establishment of desirable species which converts primary energy into fish flesh at a more economic rate and in the reservoir. Any lapse in this important management measure might lead to proliferation of undesirable species like minnows, which, in turn, become forage to predators, thus establishing a long food chain. There are many instances of establishment of such long food chains in India (Nagarjunasagar, Tungabhadra and Hirakud) which is difficult to reverse. These reservoirs harbor good standing crops of plankton and benthos, which are poorly converted into fish flesh.

However, naturalization of inducted species is quite often beset with many problems unless the species are selected with care. It needs to be ensured that the species stocked should find the habitat conductive to their biological and physiological requirements, they have an edge in the competition for food and finally, the environmental conditions favour due to the erratic hydrographic conditions that break the breeding rhythm have been found to be single major factor responsible for the failure of stocked fishes to hold out in reservoir.

4.7.7. Selection of Species for Stocking

The principles to be followed in the selection of species for 'stock enhancement' are summarized below:

1. The species should have a fair chance of establishing itself as a naturalized population.
2. The species should find the environment suitable for growth and reproduction.
3. It should be quick growing, ensuring high efficiency in food utilization.
4. A fishery comprising herbivoures with a short food chain is preferable, as they have a better conversion of primary production to fish flesh.
5. Existing laws and regulations should not be violated and necessary clearance need to be sought for introducing species where applicable.
6. Seed should be readily available or can be raised in large numbers near the reservoir.
7. Cost of stocking and managing the species must be such that the operation becomes economically viable.

Fish population in a newly impounded reservoir will be invariably too thin for the amount of fish food available in the system, leaving a void between carrying capacity and actual production of fishes. To fill this gap, there is a need to build a higher population density, which is achieved by addition of species to the original fish populations. In addition, informations on differences, if any, in the growth rates of the emdemic and stocked species, and the time taken by the stocked species in attaining harvestable size would provide insight into the production dynamics of the system. Both intra-and inter-specific competitions are to be considered in the stocking programme. Situations where two or more species use a similar resource such as food or space lead to overcrowding and poor growth rate.

Indian reservoirs, by large, have a wide ranging representation of biotic communities. Phytoplankton comprising Cyanophyceae, Dinophyceae and Bacillariophyceae dominate over the zooplankton such as copepods, cladocerans, rotifers and protozoans. Benthos is represented by insect larvae and nymphs, oligochaetes, nematodes and mollusks. There is a rich growth of periphyton on the submerged objects but the large magnitude of water level fluctuations does not favour the establishment of aquatic macrophytic communities. Significantly, many of the above niches with the exception of insects, Cyanophyceae and mollusks are shared between Indo-Gangetic major carps, carp minnows and weed fishes, necessitating control on the latter two groups. The ecosystem-oriented management policy places due emphasis on trophic strata in terms of shared, unshared and vacant niches. As mentioned earlier, the two main pathways through which primary energy finds its way to fish flesh are the grazing chain and the detritus chain. Based on this fact, either plankton feeding or detritus feeding fishes are considered suitable fo0r stocking Indian conditions.

Today, reservoir fisheries in India largely centre on development of carp fisheries. Their unmistakable role has been demonstrated in the Gangetic as well as peninsular reservoirs. Major carps, by virtue of their feeding habits and fast growth rate are indispensable in reservoir management. However, the Indian major carps are ill suited to utilize phytoplankton, the most dominant fraction of plankton. Development of other endemic species as stocking material has not made much headway in the country although some of them have a proven track record in ensuring an efficient energy transformation rate *P. pangasius*, subsisting on a molluscan diet is a species to be considered in the detritus-based, mollusc-rich reservoirs of the country. *Puntius pulchellus*, the peninsular species is a well-known macrophyte feeder and *Thynnichthys sandkhol* is reported to consume Microcystis, the common blue green alga in Indian waters. Diversification of stocking material is essential for establishment for establishment of a multi-species fish stock that utilizes all food niches of the ecosystem. In reservoirs, where annual draw down in not pronounced and water level fluctuations are not steep, phytobenthos and macrovegetation develop in various degree. The grass carp, *Ctenopharyngodon idella* can be considered for such water-bodies. The common carp is being stocked in many reservoirs. This sluggish fish does not survive normally in the warm, deep-basin reservoirs of the south, especially when infested with predators. However, this prolific feeder could carve out a place for itself in the reservoir of the northwest and north-east and in some of the peninsular reservoir like Krishnarajasagar.

Stock Rates

Amount of food available in the new environment has a considerable bearing in determining stocking rates and hence production. Fish production from unit area in a product of individual growth rate and population density. From a study of growth rate of various species in a particular body of water, it is possible to assess their optimal stocking density. The sizes of fingerlings that are stocked in Indian reservoirs come under the pre-recruit phase and so, up to the size of entering the exploited phase they are prone only to natural mortality. Therefore, a knowledge of the natural mortality rate is essential as due compensation can be provided for it while computing optimum stocking rates. While estimating optimum stocking rates for such population, about which no reliable estimates of natural mortality are available, it is felt that assumption of higher natural mortality rate would be desirable as a little overstocking would be less harmful than understocking.

A number of methods are in vogue for calculating the stocking rate. Essentially, the number of fingerlings to be stocked is the total biomass to be harvested divided by the biomass of individual fish. However, as all fish stocked are not harvested, some addition will be needed to compensate for the possible stock loss due to mortality. A general stocking formula, which can be applied universally, irrespective of the size of the reservoir is given below. This is the most popular formula among the fishery managers as unskilled workers without any difficulty can estimate this. The fish yield rate can be unskilled workers without any difficulty can estimate this. The fish yield rate can be estimated from the primary productivity studies. While estimating the percentage loss, chances of survival of the stocked fish in the light of predator pressure, escapement through the outsets and overfishing are to be taken into consideration:

$$\text{Stocking rate (no/ha)} = \frac{\text{Total yield (kg/ha) + loss (per cent)}}{\text{Individual growth rate of fish (kg)}}$$

Fish Stock Monitoring

Experience in a number of medium and large reservoirs prompts us to conclude that stocking can be considered as successful only if the stocked fish survive, grow, breed regularly and their natural breeding contributes to effective auto-stocking. In some cases, despite persistent stocking, the transplanted species have failed to show up in a catch, thereby rendering the expenditure incurred in stocking as waste. This is primarily due to stock monitoring/management measures. It is essential to ensure that the stocked fishes grow to maturity and breed naturally and the juvenile achieve desired levels of survival so that the natural recruitment takes place. A number of measures are required for this. A standard stock assessment method needs to be used to monitor the population dynamics of the reservoir to monitor the stock. There should be restriction on fishing till the fish stocked in the first few years mature and breed in the reservoir. Other steps needed are:

1. Protection of brood stock,
2. Protection of breeding grounds,
3. Protection of feeding grounds of juveniles, and
4. Enforcement of mesh regulations to prevent catching of young ones. Despite all this, there might be some reservoirs where stocked species will fail to establish.

Initial Stocking

It is very essential to continuously monitor the fish stock in the reservoir to make appropriate decisions. Standard population dynamics models are available for this purpose. During the initial 2-3 years, there should be heavy stocking and restricted fishing activity to allow the stocked fish to grow to adult size and breed.

During this period, gear that target stocked fishes should be banned altogether, but the gear for predators and other uneconomic species can be allowed.

Protection of Breeding and Feeding Grounds

Fish species have diverse breeding habits. The Indian major carps are known to migrate to the breeding grounds situated upstream. Tilapia makes nest at the littoral zone and common carp breeds on aquatic vegetation. Indian major carps migrate to the breeding grounds upstream through narrow shallow streams they become easy prey for fishers. These brooders need to be protected while moving to and at the breeding grounds to allow them to complete the breeding. Similarly, the juveniles flock the shallow areas and fishing should not be in such places. In stock enhancement, the breeding population is established during the early phase of management. However, need for stocking might arise whenever there is a breeding failure or when stock is lost due to monsoon failure, flooding of spillways *etc.*

The following are the essential aspects of successful enhanced capture fisheries:

1. The right species should be selected for stocking.
2. Stocking should be done heavily for a few years during the early phase of reservoir formation (during trophic burst stage)
3. Fishing should be restricted during the first few years after stocking to allow the stocked fish to mature.
4. Breeding and feeding grounds should be protected.
5. Catching of broodstock and juveniles should be prevented through appropriate measures.
6. Stock assessment should be done through standard population dynamics modeling.
7. Supportive restocking should be done when breeding failure or stock loss occurs.
8. The community should be organized into highly motivated and empowered group so that they can manage the resource effectively.
9. Necessary marketing and financing channels should be provided in order to protect the fishers from unscrupulous money lenders and middlemen.

4.7.8. Aspects of Reservoir Fisheries in India

Other aspects of fish stock monitoring are discussed here.

Seed Production Infrastructure

Stocking is the key to success in both culture based as well as stock enhancement regimes. Most of the instances of low yield from reservoirs in India can be attributed to non-compliance of stocking size and numbers. The established carp seed industry in the country caters largely to the aquaculture sector and the large demand from reservoirs often remains unmet. This calls for creating an adequate supply chain for fingerlings for stocking in reservoirs. In areas where large numbers of small reservoirs are taken up for development, a cluster approach would be ideal and also cost-effective. Whenever land-based nurseries are not available, pea and cage culture for in-situ raising of seed material would be most desirable.

Unplanned Stocking

The Indian experience of stock enhancement has been very revealing. The Indian rivers are known for their rich faunistic diversity and many of the native species are considered to be superior to the introduced ones in their growth performance. Still, almost all reservoirs are routinely stocked with Indian major carps. while there are some instances of successful transplantations where the stocked species established good breeding populations in the reservoirs leading to increased yield rates, in many cases, stocking attempts were rarely governed by any ecosystem considerations. These irrational stocking efforts were not only wasteful exercises, but they also proved to be detrimental to the native fish stocks from the conservation and yield optimization points of view. The policies adopted in Indian reservoirs mainly consist of stocking fingerlings of a species or a combination of species without giving any consideration for the purpose of stocking. No criteria are followed in respect of stocking rates, species mix based on the biogenic capacity of the reservoir and without any follow-up to achieve establishment of naturalized populations. On many occasions, the main consideration for rate of stocking and the species selection is the availability of fingerlings.

Crafts and Gear

Crafts

Coracle, a saucer shaped country craft, is the major fishing craft used in the reservoirs of peninsular India. It is made of a split bamboo frame, covered with buffalo hide. Apart from being simple and inexpensive, coracle is durable and has very good manoeuvrability in choppy waters. It is also a versatile craft used for laying and lifting of nets, besides navigation and transport of fish and other material. Coracles of Krishnarajasagar are prepared by wrapping HDPP over the bamboo frame with the help of coal tar as an external covering in place of hide. This modified version of coracle is cheaper and more durable.

Unlike Gobindsagar, where all the fisherman posses their boats, reservoir fisherman, in general, are too poor to own boats. In many reservoirs like Vallabhasagar and Hirkud, the fisherman could get their boats with the help of subsidy and other financial assistancre from the Government or funding agencies. In Vallabhasagar, boats are distributed by the state among the fisherman at a subsidy of 50 per cent while in Hirakud, they get it from various schemes under NABARD and NCDC. Wooden boats are used for fishing in a number of reservoirs, especially in north India. Flat bottomed, locally fabricated boats ranging in length from 3 m to 7 m are used in Kyrdemkulai, Hirakud, Malampuzha, Gobindsagar, and Rihand.

A plank-built, flat-bottomed canoe, 2m to 3m in length is the most popular fishing craft of Gandhisagar. In the same reservoir, the repatriates from the erstwhile East Pakistan used the Bengal type dinghy, which is 5m to 7m in length and have the additional facility of setting sails for wind propulsion.

Mechanized boats are not used in reservoir fishing in any appreciable extent. A 9.1 m long wooden, mechanized boat has been introduced by the Central Institute of Fisheries Technology (CIFT) in Hirakud reservoir, but they are too expensive for the fisherman. It is significant to note that large water bodies like Nagarjunasagar, Tungabhadraand Krishnarajasagar have no motorized craft neither for fishing nor for fish transport.

Dugout canoes, carved out of plan trees are used in Yerrakalava reservoir. In most of the reservoirs in the country the fisherman rely on improvised materials. Reservoir fisherman show considerable ingenuity in fabricating makeshift rafts out of discarded old tyres,logs, used cans, *etc.* In a vast majority of Indian reservoirs, where the catch is not remunerative, no boats are used and the fisherman depend entirely on these improvised devices.

Gear

The presence of underwater obstacles restricts the use of active gear in reservoirs and the choice is often limited to passive gear as simple gill nets. The most common among them is the Rangoon net, an entangling type of gill net without a footrope. Another entangling type of net used in reservoirs is uduvalai, which has a reduced fishing height and is usually operated in shallow marginal areas to catch small fish. Shore seines of various dimensions and mesh sizes are employed in many reservoirs. Although a number of other fishing gear such as long lines, ahnd lines, pole and line, cast nets, dip nets, *etc.* are in use, their contribution to the total catch is very insignificant.

Unlike the marine fisheries, very little attention has been paid over the towards improvement of gear in the inland sector, baring an attempt to upgrade the reservoir fishing gear by two experts under the aegis of FAO/UNDP programme during 1960s. A number of improvement have been suggested by the experts to the fishing techniques followed by the reservoir fisherman. Apart from the introduction of frame net, they have suggested improvements on the design of gill nets, beach seines and long lines. Unconventional methods such as electrical fishing, use of light as fish lure, the use of echo-sounder for fish detection and survey of bottom topography were also suggested. The main emphasis of gear improvement was the modifications of gill nets. Russian experts designed two nets, *viz.* Sebgul I and Sebgul II, which were gill nets with modified rigging pattern. While ensuring a proper fixing of webbing on head and foot ropes, a uniform hanging coefficient was ensured. Similarly, the sideways movement of the webbing was checked to maintain effective area of the net. Experiments were conducted in Bhavanisagar reservoir to check the efficacy of the new design and no appreciable superiority was found for Sebgul nets over the Rangoon nets.

The CIFT, Kochi has experimented with gill nets of various colours and found yellow and orange colored nets yielding better catch than the white coloured ones. It has been observed that 77 per cent of the carps were caught by entangling and the rest by gilling. The usual method of increasing the entangling capacity of gill net is by decreasing the slackness of webbing which can be achieved by suitable modifications in the hanging of the net. Gill net being a highly selective removal of carps by the gill nets can result in situations leading to domination of carp minnows, predators and other undesirable fish.

Trawling

Trawling has been attempted only in two reservoirs, *viz.* Gandhisagar and Hirakud. The findings with regard to species selectively of different type of travelling indicate efficacy of various types of trawling in increasing the productivity of commercial carps and checking the predator and weed fish populations. After experimental trawling with single-boat bottom trawling, two-boat bottom trawling and two-boat mid-water trawling under different speeds, it has been found that more than 92 per cent of the total catch consisted of economic species of fishes such as catla, rohu, murrels, mullets and featherbacks in two-boat mid-water trawling. This was in sharp contrast to the bottom trawling, of both single boat and two boat variety, which yielded mostly (64 to 91 per cent) the non-commercial species of fish. The two-boat mid-water trawling at a speed of 3 to 4 knots have been recommended for exploitation of commercial species and single and two-boat bottom trawling 2 to 3 knots for eradication of uneconomic species of fishes.

Unconventional Production Systems

Cage and Pen Culture

Production systems, such as cage and pen cultures are becoming increasingly popular in India, although they have a definite role to play in augmenting fish production form open water, especially the reservoirs. It is now widely accepted that the pen enclosures erected in the reservoir margins can be used as nurseries to raise stocking material to obviate the necessity for constructing land-based nursery farms, which are cost-intensive. Similarly, the rearing of fish in pens up to marketable size enables easier stock manipulation and total harvesting. In case of some hardy species such as common carp and tilapia, cages can also be used for raising stocking material. However, non-standardization of farm practices and the materials to be used in cage and pen culture still acts as a major retardant for large-scale adoption of these culture systems in Indian reservoirs.

Species Selection

Main criteria for the choice of candidate species for cage and pen culture are discussed here:

1. Fast growth rate.
2. Adaptability to the stresses in enclosures due to crowded conditions,

3. Ready acceptance of artificial feeds consisting mainy of cheap agricultural byproducts.
4. High feed conversion rates,
5. Resistance to disease and
6. Good market demand.

Under the Indian conditions, the Gangetic major carps (*C. catla, L. rohita, C. mrigala*), murrels the Chinese carps (*Hypophthalmichthys moltrix, Ctenopharyngodon idella*) and common carp (*Cyprius carpio*) satisfy these requirements to a great extent. Selection of species, however, is mainly dictated by the local demands and availability of quality seed and other inputs in adequate quantities.

Site Selection

Appropriate site selection is important for successful use of pens and cages. Sheletered, weed-free, shallow bays are the ideal locations for installing pens and cages. The sites should have adequate circulation of water with wind and wave action within moderate limits. Water level fluctuation is the most important consideration in site selection for the pen culture operations in reservoirs. A scrutiny of the contour map and the monthly fluctuation patterns of reservoir levels will enable the location of suitable sites which retain sufficient water for the required period of time. Sites which dry out drying summer will be ideal, as it is easier to erect pens of dry land, to be inundated later as the water for sufficient period can be identified and cordoned off by erecting barricades.

Floating fish cages can be constructed out of a variety of materials including metal, wood, bamboo and netting. Fairly fine-meshed nylon netting can be used for nursery purposes. Cages made of monofilament woven material of 1.0 mm to 3.0 mm mesh size are light and easy to handle but these last only for six months to one year, depending on their thickness. Knotless nylon webbing of 3mm to 6mm mesh size and knotted nylon webbing of 7mm to 15 mm mesh have been found to be very durable as cage material. A battery of cages can be buoyed up within a bomboo catwalk, which will serve as a working platform, floated by sealed empty barrels. Circular and box-like cages of varying dimensions on conduit pipe structures, which can be easily assembled, and suitable flotation systems have been designed in India. Similarly, self-floating cage with HDPP pipe structure has also been experimented with success.

Pen culture has a special relevance in reservoir management, since it has been widely recognized as a means to rear, in situ, the fingerlings for stocking. The number of fingerlings required for stocking the reservoirs in the country is so enormous that it is impossible to raise all of them in land-based nursery farms, which makes pen nurseries sine qua non for reservoir management. Nevertheless, pen culture on a regular basis has not been practiced anywhere in India except at

Tungabhadra reservoir. The factors that hamper the standardization of pen culture technique are:

1. The steep level fluctuations,
2. Wind and wave action,
3. Lack of suitable pen materials,
4. Weed infestation and the related harvesting problems, and
5. Non-synchronization of suitable water levels and the spawn availability.

The water retention time is important, since the rearing has to be completed before the water level in the pen goes down the critical limit. In reservoirs with high draw down, the water retention time is very limited. Sometimes the filling takes place so late that no spawn of desirable carps will be available when the water level attains the desirable limit. Recent advances made by research institutions to breed fish for protracted time give hope for solving this problem. The pen walls limiting the water circulation to some extent, the accumulated feed fertilizers cause eutrophication leading to weed infestation fouling of water and fish kills.

Other Management Measures

Pre-impoundment Surveys

In India, pre-important surveys conducted during dam construction invariably lack a fisheries perspective. However, in case of river Narmada, the Narmada Control Authority has conducted a socio-economic surveys of the Narmada basin with the objectives of addressing fisheries development in the reservoirs created by the impoundment of the river and its tributaries.

The pre-impoundment surveys help providing a framework for future fisheries and other development activities in the reservoir, which inter alia encompass the following aspects:

1. The native ichthyofauna in the river stretch above and below the dam, and their likely chances of survival.
2. Breeding habits of fishes and the possible impact of impoundment on their recruitment.
3. Survey of breeding grounds in relation to submergence, both above and below the dam.
4. Hydro-biological characteristics of water and soil with special emphasis on the nutrient and thermal regime.
5. Assessing the needs for creating infrastructure such as, hatcheries, nurseries, ice plants, *etc.*
6. Site selection for pen nurseries, cages, *etc.*
7. Possibilities for cleaning the area of submergence of trees and other obstruction.

8. Inventory of fishing villages, together with demographic details on fisher population and fishing craft and gear.

The following account deals with some of the important pre-impoundment activities:

4.7.9. Stocking

Inducting fast-growing extraneous species into the ecosystem to colonise the diverse niches is a necessary prerequisite of reservoir management. Since one of the primary aims of stocking is to ensure utilization of the enhanced food reserves, the ideal time to stock new species is the period of trophic burst. Any lapse in this important management measure causes the proliferation of trash fishes by taking advantage of the increased availability of fish food organisms, which in turn, may provide forage base for catfishes. Nagarjunasagar, Tungabhadra, Hirakud and a number of other large reservoirs in India are examples where the minnows, catfishes, murrels and other uneconomic fishes gained grounds in the early years, leading to establishment of long food chains. These reservoirs harbour good standing crops of plankton and benthos, which are not reflected in the fish output. Even intensive stocking at a later stage has failed to reverse the situation.

Since large and medium reservoirs are to be developed on the principles of capture fisheries, it is desirable to stock the species that may breed and ultimately get naturalised in the system through autostocking. This is imperative to meet the long-term objective of obtaining a sustained yield rate. Management involving persistent stocking in large water bodies not only pushes up the input cost, such systems also create many practical difficulties in raising the stocking material in adequate quantities. However, naturalisation of introduced species is quite often beset with many problems. The alien species should find the habitat conducive to its biological and physiological requirements. It should have an edge in the competition for food and finally, the environmental conditions should favour its requirements of spawning and larval development. Recruitment failure due to the erratic hydrographic conditions that break the breeding rhythm have been found to be the single major factor responsible for the failure of stocked fishes to hold out in reservoir.

In Konar, strong currents wash down the carp eggs into the deep zones of the reservoir leading to their destruction and resultant recruitment failure (Parameswaran *et al.,* 1969). But more often, the fishes do not find suitable spawning grounds as in the case of many south Indian reservoirs. Since the breeding of major carps is governed to a great extent by the magnitude of monsoon floods, annual variations in this parameter affect their breeding and recruitment.

Selection of Species for Stocking

Jhingran (1988) has summarized the principles to be followed in the selection of species for stocking as:

Table 16: The Broad Distinguishing Features of Small and Large Reservoirs

Small Reservoirs	*Large Reservoirs*
Single-purpose reservoirs mostly for irrigation.	Multi-purpose reservoirs for flood-minor control, hydro-electric generation, large-scale irrigation, *etc.*
Dams neither elaborate nor very expensive. Built of earth, stone and masonry work on small seasonal streams.	Dams elaborate, built with precise engineering skill on perennial or long seasonal rivers. Built of cement, concrete or stone.
Shallow, biologically more productive per unit area. Aquatic plants common in perennial reservoirs but scanty in seasonal ones.	Deep, biologically less productive per unit area. Usually free of aquatic plants. Subjected to heavy drawdowns.
May dry up completely in summer. Notable changes in the water regime.	Do not dry up completely. Changes in water regime slow. Maintain a conservation-pool level (= dead storage).
Sheltered areas absent.	Sheltered areas by way of embayments, coves, *etc.* present.
Shoreline not very irregular. Littoral areas with a gentle slope.	Shoreline more irregular. Littoral areas mostly steep.
Oxygen mostly derived from photosynthesis in these shallow, non-stratified reservoirs, lacking significant wave action.	Although photosynthesis is a source of dissolved oxygen, the process is confined to a certain region delimited by vertical range of transmission of light (euphotic zone). Oxygen also derived from significant wave action.
Provided with concrete or stone spillway, the type and size of the structure depending on the size of the runoff.	Provided with more complex engineering devices.
Major carps breed in the reservoirs.	Breeding mostly observed in the headwaters or in other suitable areas of the reservoir.
Can be subjected to experimental manipulations for testing various ecosystem responses to environmental modifications.	Cannot be subjected to experimental manipulations.
Trophic depression phase can be avoided through chemical treatment and draining. Cycle of fish production can be repeated as often as the reservoir is drained.	Trophic depression phase sets in.
The annual flooding during rainy season may be compared to overflowing of floodplains. Inundation of dry land results in a release of nutrients into the reservoir when it fills up, resulting in high production of fish food through decomposition of organic matter, predominantly of plant origin, leading to higher fish growth and survival.	Loss of nutrients occurs as they are locked up in bottom sediments. Rapid sedimentation will reduce benthos production.
Through complete fishing or overfishing of seasonal reservoirs, no brood stock is left. Fish stock has to be rebuilt through through stocking. There is thus established a direct relationship between stocking rate and catch per unit of effort.	Prominent annual fluctuations in recruitment occur and balancing of stock number against natural mortality requires high density stocking of fingerlings. Their capture requires efficient capture methods.

Jhingran, 1988.

1. The species should find the environment suitable for growth and reproduction.
2. It should be quick growing, ensuring high efficiency in food utilization.
3. A fishery comprising herbivores with a short food chain is preferable, as they have a better conversion of primary production to fish flesh.
4. The stocking density should be such that the food resources of the ecosystem are fully utilized and optimum population maintained, consistent with normal growth.
5. The size of the fingerlings to be stocked should be so chosen to get the desired results.
6. Seed should be readily available with minimal transportation cost.
7. Cost of stocking and managing the species must be such that the operation is economically viable.

One of the important considerations is to know the amount of food available in the new environment. This factor has a considerable bearing in determining stocking rates and hence production. Fish production from unit area is a product of individual growth rate and population density. From a study of growth rate of various species in a particular body of water, it is possible to assess their optimal stocking density. Efficient utilization of fish food communities increases the carrying capacity and therefore calls for a higher population density which is achieved by addition of species to the original fish populations. In addition, information on differences, if any, in the growth rates of the endemic and introduced species, and the time taken by the introduced species in attaining harvestable size would provide insight into the production dynamics of the system.

Competition

Both intra- and inter-specific competitions are to be considered in the stocking programme. Situations where two or more species use a similar resource, such as food or space, lead to overcrowding and poor growth rate. Under a higher than optimum stocking rate, though production may be high, the individual growth rate will be so small that to attain a marketable size a long growing period will be needed. On the other hand, if bigger fishes are needed, the rate of stocking should be lowered and a low production will have to be accepted. Similarly, when a marketable size is to be attained in a shorter period, stocking rate will have to be lowered to allow faster growth. Thus, a desired balance among stocking rate, population density and growth is to be maintained with enough flexibility so as to swing it to suit the changes in environmental factors. Such a plan must determine tentative stocking rates and population thinning, depending on the need (Jhingran, 1988).

The sizes of fingerlings that are stocked in Indian reservoirs come under the pre-recurit phase and so, up to the size of entering the exploited phase they are prone only to natural mortality. Therefore, a knowledge of the natural mortality rate is essential as due compensation can be provided for it while computing optimum stocking rates. Jhingran and Natarajan (1969), while evaluating the stocking rates of Damodar Valley Corporation (DVC) reservoirs, arbitrarily recommended a compensation factor for natural mortality @ 25 per cent for reservoirs with no large predaceous fishes and 50 per cent for those harbouring large predators. While estimating optimum stocking rates for such populations, about which no reliable estimates of natural mortality are available, it is felt that assumption of higher natural mortality rate would be desirable as a little overstocking would be less harmful than understocking. Fish Seed Committee of the Government of India (1966) recommended the stocking rate for reservoirs at about 500 fingerlings ha^{-1} of the size range 40 to 150 mm.

Stocking Measures Adopted in Indian Reservoirs

The policies hitherto adopted in Indian reservoirs mainly consisted of stocking fingerlings of a species or a combination of species without any definite density levels or ratios based on the biogenic capacity of the reservoir. Rate of stocking and the species-mix are often determined by their availability, as can be seen from the case studies in the chapters ahead.

Basic productivity of the reservoir is dependent on the amount of solar energy available and the efficiency of the system to transform it into chemical energy. Besides, the energy conversion efficiency at trophic levels of consumers differs considerably from one reservoir to another, depending on the qualitative and quantitative variations in the biotic communities. Any conversion rate above 1 per cent can be considered as good. In an ideal situation, the commercial species share the ecological niches in such a way that trophic resources are utilised to the optimum. At the same time, the fishes should belong to short food chain in order to allow maximum efficiency in converting the primary food resources into harvestable materials. But in reservoirs, such conditions seldom prevail.

Indian reservoirs, by and large, have a wide ranging representation of biotic communities. Phytoplankton comprising Cyanophyceae, Chlorophyceae, Dinophyceae and Bacillariophyceae dominate over the zooplankton such as copepods, cladocerans, rotifers and protozoans. Benthos is represented by insect larvae and nymphs, oligochaetes, nematodes and molluscs. There is a rich growth of periphyton on the submerged objects. The large magnitude of water level fluctuations does not favour the establishment of aquatic macrophytic communities. Significantly, many of the above niches with the exception of insects, Cyanophyceae and molluscs are shared between Indo-Gangetic major carps and trash fishes, focussing the need for controlling carp minnows and weed fishes. The ecosystem-oriented management policy places due emphasis on trophic strata in terms of shared, unshared and vacant niches. Two main pathways through which primary energy finds its way to fish flesh are the grazing chain and the detritus chain. Contribution by both the pathways to the total availability of the energy needs to be assessed for determining the species combination, most suited to the ecosystem. A large number of Indian reservoirs exhibit the detritus chain of energy transfer.

Prior to the development of carp seed production technology in India, natural spawn collected from rivers were stocked in reservoirs. Thus, the seed of *Puntius* spp., *Cirrhinus* spp. and *Labeo* spp. collected from the Cauvery were extensively stocked in the reservoirs of Tamil Nadu along with euryhaline species such as *Chanos chanos, Etroplus suratensis* and *Megalops cyprinoides. Labeo fimbriatus, Cirrhinus cirrhosa*, tilapia and *Etroplus suratensis* were the common stocking material in Kerala during the early years. With the advent of induced breeding, most of the States in India were able to raise the carp seed in large numbers by the 1970s and this resulted in a shift in species-mix in favour of catla, rohu and mrigal. Today, reservoir fisheries in India largely centre on development of carp fisheries. Their unmistakable role has been demonstrasted in the Gangetic as well as peninsular reservoirs. Major carps, by virtue of their feeding habits and fast growth rate are indispensable in reservoir management. However, the Indian major carps are ill-suited to utilize phytoplankton, the most dominant fraction of plankton. The remarkable ability of silver carp in efficient conversion of phytoplankton into fish flesh has been demonstrated in Kulagarhi and Getalsud reservoirs, despite the persisting doubts about the digestibility of *Microcystis*. However, introduction of exotic fishes in open waters is still a subject of controversy due to its possible deleterious effects on indigenous populations.

Development of endemic candidates as stocking material has not made much headway in the country although some of them have a proven track record in ensuring an efficient energy transformation rate. *P. pangasius*, subsisting on a molluscan diet is a species to be considered in the detritus-based, mollusc-rich reservoirs of the country. *Puntius pulchellus*, the peninsular species is a well-known macrophyte feeder and *Thynnichthys sandkhol* consumes *Microcystis*, the common alga in Indian waters. Diversification of stocking material is essential for establishment of a multi-species fish stock that utilize all food niches of the ecosystem. In reservoirs, where annual drawdown is not pronounced and water level fluctuations are not steep, phytobenthos and macrovegetation develop in various degrees. The grass carp, *Ctenopharyngodon idella* can be considered for such water bodies. The common carp is being stocked in many reservoirs. This sluggish fish does not survive normally in the warm, deep-basin reservoirs of the south, especially when infested with predators. However, this prolific feeder could carve out a place for itself in the reservoirs of the north-east, in Gobindsagar and in some of the peninsular reservoir like Krishnarajasagar. The fish being a mud-stirrer, is considered to be unsuitable for already turbid waters.

Tilapia, due to its records of rapid proliferation and consequent stunted growth in pond ecosystem, does not find favour with many fishery managers, although in Amaravathy and Malampuzha it has performed well. Stocking of prawn, *Macrobrachium malcolmsonii* has been tried in Tungabhadra and Konar (Natarajan, 1979b), where they could not survive and contribute to the commercial fisheries. Ahmed (1993) emphasised the need for extensive stocking of this prawn, by quoting instances of its self-sustaining populations in many reservoirs. However, the prawn's ability to complete its full life cycle in the freshwater phase is yet to be proved. *T. putitora, L. dero*, and the exotic species such as mirror carp, silver carp, grass carp, *Tinca tinca* and *Carassius carassius* are advocated for the high altitude reservoirs (Natarajan, 1979b).

Impact of Stocking

The most important objective of stocking, *i.e.*, to augment the yield, can be achieved only if the stocked fishes survive, grow and get caught in the fishing gear. This is achieved, to a large extent, in small reservoirs where the management centres round the stocking and recapture system. However, in larger water bodies, the recapture is uncertain on account of many reasons as mentioned earlier.

Impact of Stocking in Medium and Large Reservoirs

Experience in a number of medium and large reservoirs prompts us to conclude that the stocking programme can be termed as successful, only when the stocked fishes breed in the reservoir and contribute towards autostocking. In many cases, despite persistent stocking, the transplanted species did not show up in the catch, thereby rendering the expenditure incurred in stocking as waste. Only in a few instances the resources mobilised for stocking operation were compensated by generation of income through recapture of the stocked fishes.

Sreenivasan (1984) reviewed the impact of stocking in 10 reservoirs of Tamil Nadu. The stocked catla built up a naturalised population in Mettur reservoir. Just 10,000 fingerlings were stocked during 1922 to 1935, which formed the nucleus of a self-propagating stock and dominated the catch during the 1960s. Catla fisheries, however, suffered periodic setback due to breeding failures. The current contribution is as low as 10 per cent. Recapture of two other stocked fishes *viz., L. rohita* and *L. calbasu* is reported to be adequate (Sreenivasan, 1984). However, stocking of *L. fimbriatus* (over 2 million), common carp (1 million), *L. kontius* (0.4 million), *P. carnaticus*(0.4 million), *C. reba* (several hundred thousands) and *P. dubius* (several hundred thousands) is believed to be wasteful, since they were never recaptured in any appreciable quantity.

In Bhavanisagar, although the transplanted catla showed up in the catch, they could not make any impact on the catch structure. As opposed to this, *L. calbasu*, another transplant, found the environment congenial for breeding and established itself in the reservoir, supporting a lucrative fishery for the last few decades. More than 2 million common carp fingerlings were stocked in the reservoir, of which only a few hundreds were recaptured. Intensive stocking of *L. fibriatus* did not make any dent on the fisheries, despite the fact that the fish was native to the system.

In Sathanur, a breeding population of catla has been successfully established through stocking and the fish contributes more than 80 per cent of the total catch. The species has also eclipsed the indigenous *L. fimbriatus* by reducing its contribution from 36 per cent to 1 per cent. It is pertinent to note that the stocking done over 12 years involving 2 million fingerlings of indigenous species such as

C. cirrhosa, L. kontius, C. reba and *L. fimbriatus* could not make any impact on restoration of their fisheries, primarily due to the inability of the fishes to breed and propagate themselves.

In Krishnagiri, although the increase in percentage of the stocked major carps has resulted in some initial increment in their contribution in the catch, the recapture was not commensurate with the stocking rate. Stocking of Gangetic carps, common carp and *L. fimbriatus* done in Vaigai, so far, has been described as wasteful (Sreenivasan, 1984), where presently, catla contribute 20 per cent of the catch.

In Malampuzha and Peechi, the two medium reservoirs in Kerala not register any substantial increase in yield rate despite sustained stocking with Gangetic major carps, mainly on account of their failure to breed. In Malampuzha, despite a sharp increase in stocking density from 52 ha^{-1} in the 1970s, and a corresponding increase in the percentage of major carps in the stocked fishes, the yield rate remains at a low level of 5.0 kg ha^{-1}. Although catla grows to an impressive size, the contribution of major carps never exceeded 20 per cent of the total catch. Considering the low yield rate, the quantity of major carps harvested does not commensurate the stocking effort. Similarly, in Peechi, 90 per cent of the fingerlings stocked belong to the Indian major carps, especially *L. rohita.* But they are not reflected adequately in the catch and the yield rate remains at a low level of 4.5 kg ha^{-1}.

In Nagarjunasagar, Andhra Pradesh, regular annual stocking at the rate of 50,000 to 833,000 fingerlings comprising catla, rohu and mrigal during the 1970s had little impact on the catch structure, as none of the stocked fishes could breed and contribute to recruitment. Similarly, stocking as a management option has failed in Tungabhadra reservoir situated on the same basin and Krishnarajasagar in the Cauvery basin. In Tungabhadra, the stocking rate during 17 years ranged from 1 to 11 ha^{-1}, while it varied from 0.15 to 67 ha^{-1} in Krishnarajasagar. In both the reservoirs, the selection of species has always been arbitrary, the stocking density was inadequate and the stocked fishes failed to breed in the reservoir.

Gandhisagar in Madhya Pradesh is an example of the stocked fishes contributing to fish catch in a sustained manner through breeding and recruitment. During the 1950s through 1970s, on account of steady stocking of catla (2 million), rohu (1.3 million), and mrigal (1.1 million), the fish yield rose to 20.33 kg ha^{-1} from the initial 0.51 kg ha^{-1}. Among the stocked fishes, catla contribute 60 to 70 per cent, while mrigal and rohu form only 1 to 20 per cent. Mrigal, despite being a part of the indigenous ichthyofauna, is on the decline. Ravishankarsagar in the same State was stocked, on an average, with 1.77 million fingerlings every year and yet the yield rate could not be raised above 8.3 kg ha^{-1}, primarily due to the non-establishment of the carps. The amount spent on stocking has been much higher than the cost of recaptured fishes. Rihand reservoir in U. P. could build up a breeding population out of the initial stocking. Although the yield rate of 0.58 kg $ha^{-1}$is not impressive, 73 to 99 per cent of the catch comprises *C. catla.*

The success of stocked Indian major carps in Ukai reservoir in Gujarat can also be attributed to their breeding in the reservoir. Apart from augmenting the reservoir fisheries through recruitment, the young ones of Indian major carps are also reported to escape through the outlet of the dam and contribute to stocking of downstream impoundments. In all the DVC reservoirs, *viz.*, Konar, Tilaiya, Maithon and Panchet, stocking did not have any impact (Sreenivasan, 1984). Some breeding activities in respect of catla and mrigal have been reported, though low survival rate of the spawn and frequent breeding failure due to erratic monsoon prevented proper recruitment of the planted species.

Indian experience of stocking medium and large reservoirs suggests that by and large, the stocking becomes effective only when the stocked fishes propagate themselves. Moreover, this breeding population can be built up only if the stocking is resorted to during the early phase of the reservoir formation.

Impact of Stocking in Small Reservoirs

In sharp contrast to the large and medium reservoirs, stocking has been more effective in improving the yield from small reservoirs as success in the management of small reservoirs depends more on recapturing the stocked fish rather than on their building up a breeding population. The smaller water bodies have the advantage of easy stock monitoring and manipulation. Thus, the smaller the reservoirs, the better are the chances of success in the stock and recapture process. In fact, an imaginative stocking and harvesting schedule is the main theme of fisheries management in small, shallow reservoirs. The basic tenets of such a system involve:

1. Selection of the right species, depending on the fish food resources available in the system.
2. Determination of a stocking density on the basis of production potential, growth and mortality rates.
3. Proper stocking and harvesting schedule including staggerred stocking and harvesting, allowing maximum growout period, taking into account the critical water levels.
4. In case of small irrigation reservoirs with open sluices the season of overflow and the possibilities of water level falling too low or completely drying up, are also to be taken into consideration.

Aliyar is a standing testimony to the efficacy of the management based on staggered stocking. The salient features of the management options adopted in Aliyar are:

1. stocking is limited to Indian major carps (earlier, all indigenous, slow-growing carps were stocked)
2. Increasing the size at stocking to 100 mm and above.

3. Reducing the stocking density to 235–300 ha^{-1} (earlier rates were erratic ranging between 500–2 500 ha^{-1}).
4. Staggering the stocking, and
5. Regulating mesh size strictly and banning the catch of Indian major carps < 1 kg in size.

A direct result of the above management practice was an increase in fish production from 1.67 kg ha^{-1} in 1964–65 to 194 kg ha^{-1} in 1985.

Effective recapture of the stocked fishes renders the stocking more remunerative in small reservoirs. Successful stocking has been reported from a number of small reservoirs in India. In Markonahalli, Karnataka, on account of stocking, the percentage of major carps has increased to 61 per cent and the yield increased to 63 kg ha^{-1}. Yields in Meenkera and Chulliar reservoirs in Kerala have increased from 9.96 to 107.7 kg ha^{-1} and 32.3 to 275.4 kg ha^{-1} respectively through sustained stocking. In Uttar Pradesh, Bachhra, Baghla, and Gulariya reservoirs registered steep increase in yield through improved management with the main accent on stocking. An important consideration in Gulariya has been to allow maximum growout period between the date of stocking and the final harvesting, *i.e.*, before the levels go below the critical mark. The possible loss due to the low size at harvest was made good by the number. Bundh Beratha in Rajasthan, stocked with 100,000 fingerlings a year (164 ha^{-1}) resulted in a fish yield of 94 kg ha^{-1}, 80 per cent of which constituting catla, rohu and mrigal (Table 17).

Table 17: High Yields Obtained in Small Reservoirs Due to Management

Reservoir	*State*	*Stocking Rate (number ha^{-1})*	*Yield (kg ha^{-1})*
Aliyar	Tamil Nadu	353	194
Meenkara	Kerala	1 226	107
Chulliar	–do–	937	316
Markonahalli	Karnataka	922	63
Gulariya	Uttar Pradesh	517	150
Bachhra	–do–	763	140
Baghla	–do–	?	102
Bundh Beratha	Rajasthan	164	94

Instances where intensive stocking of Indian major carps became ineffective in small reservoirs are very rare. In Govindgarh, despite stocking of Indian major carps at the rate of 19 to 390 fingerlings ha^{-1}, the yield remained at 15.92 kg ha^{-1}, during the '60s and the '70s. Large-scale escapement of fishes through the open weir is believed to be the main reason for this low fish yield. An estimated 1.1 million fingerlings of catla, rohu and mrigal were stocked into Badua reservoir, Bihar during the period 1975 to 1979. However, the fish yield from the reservoir during the period remained within 4 to 7 kg ha^{-1}. Other management measures taken in the reservoir are not known. Sreenivasan (1984) reported disappointing recapture of major carps after their heavy stocking in Manjalar reservoir (Tamil Nadu). Proliferation of the tilapia, *O. mossambicus* is the main factor that prevented the major carps from getting a foothold in the fishery, probably due to their competition for food.

Removal of Predators and Weed Fishes

Presence of predatory and weed fishes poses impediments in survival and growth of economic species in many Indian reservoirs. Keeping these unwanted population under check is a very difficult management problem, especially in large reservoirs. A small population of predators helps to crop the trash fishes which compete for food with the economic species. A small predator population of Gobindsagar which keeps the minnows under check is a good example. However, no scientifically sound methods are available to keep a limited population of predatory species. Repeated use of gill nets of appropriate mesh size, use of long lines, traps, *etc.* are suggested for control of the uneconomic and undesirable populations. Manipulation of reservoir level with a view to checking the breeding and destruction of the young ones of predators and the minnows has been tried in several countries. However, this is not practicable in many Indian reservoirs since water release pattern is dictated by priority sectors like irrigation and power generation. Poisoning of selected sheltered areas arms and coves as practised abroad has also limited use in India due to the multiple use of water and objections from the other water users. David and Rajagopal (1969) reported that non-selectivity of shore seines helped in reducing catfish population in Tungabhadra reservoir by 76 to 81 per cent. *Alivi*, the giant shore seine of Tungabhadra also removes the trash fish in large numbers. Judicious use of this gear, with a condition that the juveniles of economic species are released back, can go a long way in containing the trash fish population.

Recent findings of Kartha and Rao (1990) with regard to the efficacy of trawling in checking predators and trash fishes are of interest. Bottom trawling in Gandhisagar was found to catch 64 to 91 per cent of the unwanted fishes and this has been recommended as a method to crop the predators and carp minnows. Nevertheless, applicability of this method is limited to places where the bottom is free from obstructions. Natarajan (1979b) suggested biological control of trash fishes by stocking two euryhaline species *viz., Megalops cyprinoides* and *Lates calcarifier*. Since these predatory fishes do not breed in freshwater, they cannot go out of control.

Exotic Fishes and their Role in the Reservoir Fisheries of India

In spite of an already rich and diverse fish genetic resource of India, more than 300 exotic species have been introduced into the country so far (Jhingran, 1989a). While a vast majority of them are ornamental fishes which remain, more or less, confined to the aquaria, some others have been introduced in aquaculture and open water systems with varying degrees of success. Three larvicidal fishes *viz.,*

Lebistes reticulatus, Nothobranchus sp. and *Gambusia affinis* were introduced for containing the insect larvae in confined waters. Silver carp and the three varieties of common carp were brought into the country with the objectives of broadening the species spectrum in aquaculture and increasing the yields through better utilization of trophic niches. In recent years, the bighead carp *Hypophthalmichthys nobilis* and *O. niloticus* have been reported from the culture systems of eastern India. After unauthorised introduction, these two fishes are becoming popular among the aquaculturists of the region. While a few of the introduced species proved to be a boon in aquaculture and acted as an instrument for yield optimisation from ponds, the accidental and deliberate introduction of some of the exotic fishes into the open-waters has generated a lot of debate in recent years. There is a growing concern in India about the possible deleterious impact of the exotic fishes on the fish species diversity of the Indian rivers.

Oreochromis mossambicus, Hypophthalmichthys molitrix, Ctenopharyngodon idella, Cyprinus carpio communis, C. carpio specularis and *C. carpio nudus* have gained entry into the reservoir ecosystem through accidental or deliberate introduction. Among them, tilapia, silver carp and common carp could make a negative impact on the fisheries in various reservoirs in the country. Instances of *Gambusia affinis* getting naturalised in reservoirs are rare. In Markonahalli reservoir, the fish has established itself as a breeding population and reported to be affecting the larval stages of commercially important fishes.

Tilapia

The tilapia, *O. mossambicus* was first introduced into the pond ecosystem of the country in 1952 and soon it was stocked in the reservoirs of south India. By the end of 1960s, most of the reservoirs in Tamil Nadu and those in the Palakkad and Trissur districts of Kerala were regularly stocked with tilapia. Performance of tilapia in ponds of south India has been discouraging mainly due to its early maturity, continuous breeding, over-population and dwarfing. It is reported to mature at 6 cm length at an age of 75 days and to breed at an interval of one month under the tropical conditions.

Performance of Tilapia in Reservoirs

The warm waters of the tropical reservoirs in India have provided a conducive habitat for the tilapia and it has established a secure position in a number of south Indian reservoirs. The fears of its stunted growth have been allayed as the average size of tilapia did not decline as much as it did in ponds. Sreenivasan (1967) stated that the fluctuating water levels affected the breeding pits of the fishes and the predators took a heavy toll of their young ones. These two factors are believed to keep check on the excessive proliferation of tilapia in reservoirs.

Size of tilapia in the commercial catches of reservoirs has been very good, as opposed to the unmarketable size reported fromthe ponds. The average size of tilapia from Tamil Nadu reservoirs has been 1.5 kg during the 1960s, with the minimum size of 500 g. Similarly, tilapia weighing 2.5 kg was very common in Malampuzha reservoir, Kerala during the 1960s, with an average size of 1.5 to 1.75 kg. The present size of 0.5 to 0.7 kg in Malampuzha and 0.68 kg in Tamil Nadu reservoirs are well within the limits of market preference, their continuous slide in size over the years is a cause of concern as it is feared that if the fall in size continues, it may become unmarketable.

Tilapia has dominated and virtually eliminated all other fishes including the stocked Gangetic carps in a number of reservoirs in Tamil Nadu. Vaigai, Krishnagiri, Amaravathy, Uppar and Pambar reservoirs in Tamil Nadu are harbouring sizeable populations of tilapia since 1960s, contributing substantially to commercial catches. While its contribution has declined since 1979–80 in Vaigai, it continues to form a major fishery in all the other reservoirs. In Krishnagiri, the fish has a changing fortune on account of competition with the mullet, *Rhinomugil corsula*. From the predominant position in the 1960s the percentage of tilapia came down to 4.3 per cent in 1983–86, only to increase in the year 1989–90 to 69 per cent (Jhingran, 1991). At present, tilapia forms 24 per cent of the catch.

Chulliar, Meenkara, Peechi and Malampuzha reservoirs in Kerala have been stocked with tilapia in the early sixties and the fish contributes substantially to the catch. In the relatively large (2 313 ha) Malampuzha reservoir, tilapia is reported to have registered a fast growth rate and large size, contributing 10 to 70 per cent of the catch in different years. Tilapia also found its way into Kolleru lake in Andhra Pradesh and Sondur reservoir in Raipur district, Madhya Pradesh. Jhingran (1991) examined the yield of tilapia in different classes of reservoirs in the size range 50 to 10,000 ha and came to the conclusion that, by and large, small reservoirs in the size range of 50 to 200 ha showed a better yield (44 to 101 kg ha^{-1}), than the larger ones. Relatively large reservoirs such as Malampuzha and Amaravathy have a better size of tilapia, although documentation is not sufficient to attempt any correlation between the size attained by the fish and the area of reservoirs.

Distribution of tilapia is more or less restricted to the tropical belt as the fish is constrained with slow growth and winter mortalities in the higher latitude. Attempts to introduce the fish in Baghla, a small irrigation impoundment in the Gangetic plain have not succeeded (Jhingran, 1991). The ongoing debate on the introduction of tilapia centres round its:

1. Suitability to enhance yield through niche utilization,
2. Propensity for affecting or even replacing the native ichthyofauna, and consumer preference.

One of the main considerations in determining introduction of exotic fishes is their feeding habits. Since none of the Indian culturable carps feeds on Cyanophyceae blooms like *Microcystis aeruginosa*, tilapia is often cited as a welcome addition to the blue-greens-dominated water bodies. Although Sreenivasan (1967) and many workers abroad expressed their reservations about the ability of tilapia to digest and assimilate *Microcystis*, Jhingran (1991) believed that the sodium-calcium ratio in alkaline reservoirs broke down the cell walls of blue-greens, facilitating their digestion by tilapia. *Oreochromis mossambicus* has wider omnivorous food spectrum, compared to many other species of tilapia.

Tilapia versus Indigenous Carps

The apprehensions about tilapia affecting the native ichthyofauna seem to be valid, going by its track record in India and abroad. In a number of reservoirs in India, introduction of tilapia has resulted in poor growth rate or even elimination of the indigenous species and the transplanted Gangetic carps. Sreenivasan (1967) found that the growth rates of *C. catla, L. fimbriatus* and *C. mrigala* were adversely affected by tilapia in Ayyamkulam pond. He also observed that growth of *Chanos chanos* was restricted to less than 100 g yr^{-1}, against the usual 500 g yr^{-1} in many water bodies of Tamil Nadu due to its co-existence with tilapia. In Kabini reservoir tilapia has adversely affected the indigenous *Cirrhinus reba*. During the period from 1980–81 to 1984–85, tilapia has caused decrease of *C. reba's* share in the catch from 70 per cent to 20 per cent (Murthy *et al.,* 1986). Tilapia and the Indian economic species sharing a common food niche, the success of one in competition with the other is determined by the ability to breed and propagate. Given the propensities of tilapia for autostocking, the indigenous species, which are prone to breeding failure, have a definite disadvantage in its struggle to coexist with the former.

Introduction of *O. mossambicus* and *Tilapia zillii* for weed and insect control in Californian reservoirs has affected the native ichthyofauna (Moyle, 1976). A more devastating effect on indigenous fish fauna has been reported from Kyle reservoir in Zambia, where a valuable local species *Paretopus petite* has been eliminated from the ecosystem (Lamarque *et al.,* 1975).

Consumer Preference

Tilapia of right size has a good consumer preference. Tilapia is also known as a species affordable by the poor. In Palakkad and Trissur districts of Kerala, where the fish is an important component of reservoir catch, tilapia enjoys a consumer preference over the Indian carps even when sold at an equal price (Jhingran, 1991).

Relevance of Tilapia in Reservoir Fisheries

Tilapia (*O. mossambicus*) has entered the Indian scene, when the inland fisheries contributed negligibly to the total fish production in the country and the ecosystem management was in its infancy. Today, a number of indigenous species are available for stocking to broaden the species spectrum, bridge the gaps in niche utilization and increase the yield. Barring a very few reservoirs, tilapia-dominated fishery invariably leads to low yields. In many reservoirs like Krishnagiri and Vaigai, the production has been found to be erratic due to the unpredictive behaviour of tilapia population due to competition from other fishes. Fishery managers of India are striving hard to change the fish from its dominant position, wherever, they occupy one. Thus, in the present context, tilapia does not figure among the species preferred for stocking in Indian reservoirs.

Oreochromis niloticus has not yet entered the reservoir ecosystem in India. Confined to the estuarine and freshwater wetlands of the eastern India, the fish has registered an impressive growth of 250 g in 6 months. Since this fish is not reported to have problems of stunted growth and prolific breeding, it may probably have a more positive role to play in Indian reservoirs, compared to *O. mossambicus*.

Silver Carp

Silver carp, *Hypophthalmichthys molitrix* was introduced in India in 1959 and unlike tilapia, it has not strayed into many reservoirs. However, silver carp has attracted more attention from the ecologists and fishery managers, generating a more animated debate. Importance of silver carp in reservoirs emanates mainly from:

1. Its reported ability to utilise *Microcystis*
2. The impressive growth rate, and
3. Its propensities for affecting the indigenous species, especially *Catla catla*.

The silver carp has a specialised structure of gill rakers adapted to microplankton feeding (Inaba and Nomura, 1956). Gut analysis of the fish carried out at the Central Inland Fisheries Research Institute revealed a wide feeding range including Chlorophyceae, Cyanophyceae, Chrysophyceae, Bacillariophyceae, Dinophyceae, Protozoa, Rotifera, Cladocera, Ostracoda, and Copepoda (Jhingran and Natarajan, 1978).

Performance of Silver Carp in Reservoirs

An experimental consignment of 239 fingerlings of silver carp was stocked in Kulgarhi reservoir (Madhya Pradesh) in 1969. Based on the recapture of 8 specimens, growth rates ranging from 597 mm in 783 days (0.76 mm day^{-1}) to 404 mm in 293 days (1.4 mm day^{-1}) have been recorded (Rao and Dwivedi, 1972). The fish was also introduced in Getalsud reservoir, Bihar in 1974, where it has recorded growth rates ranging from 2.20 to 5.79 g day^{-1}. In both reservoirs, the fish did not breed. There are reports about the stocking of silver carp in a number of reservoirs across the country ranging from Gumti in the north-east to Aliyar in Tamil Nadu. However, it did not get established as a breeding population anywhere.

The most spectacular performance of silver carp has been reported from Gobindsagar reservoir, where after an accidental introduction, the fish formed a breeding population and brought about a phenomenal increase in fish yields. (see the chapter on Himachal Pradesh). Silver carp is instrumental in enhancing the fish production from the reservoir from 160 t in 1970–71 to 964 t in 1992–93. Many workers have suggested the over-intensity of feeding in respect of silver carp, as reflected by the *gorged* and *full* conditions of gut all through the year. Jhingran and Natarajan (1978) pointed out that the silver carp, being a cold-water fish, when introduced into a warm regime, consumed food much in excess and grew faster as expected of a true poikilotherm. A similar latitude- induced change worth noticing is the age at maturity. The fish mature when they are 5 to 6 years old in North China, 4 to 5 years in Central China and 2 to 3 years in south China. In India, it breeds just at the age of one year under optimum conditions.

Silver Carp versus Catla

Pond culture experiments conducted in the country have unmistakably established the superior performance of silver carp when cultured along with catla (Sukumaran, *et al.,* 1968). Jhingran and Natarajan (1978) expressed the view that this need not hold good for large water bodies like reservoirs. They argued that the silver carp short-circuited the food-web, resulting in the poor performance of catla. For instance, silver carp showed preference for smaller zooplankters especially rotifers and nauplii. It was quite likely that too much of grazing on copepode nauplii by silver carp could disturb the life cycle of copepodes in a small water body like pond, causing poor performance of catla. While advocating a cautious approach, they advocated stocking of silver carp in closed reservoirs like Gobindsagar and Nagarjunasagar which are blocked by dams, both down- and upstream and no connected with the Ganga river system, the original abode of *C. catla*. The fish was not stocked in either of the reservoirs. However, in Gobindsagar, stocking of silver carp has since become irrelevant as the fish has already carved out a niche for itself in the reservoir. In the process, the lurking fear that the exotic fish is deleterious to the populations of indigenous catla has been proved beyond any doubt. Ever since silver carp gained a stronghold in the reservoir, catla which constituted an annual fishery of the magnitude of 200 to 300 t has declined considerably.

Karamchandani and Mishra (1980), while evaluating the co-existence of silver carp and catla, established that the two fishes shared a common niche and compete with each other for food in a reservoir ecosystem. Percentage composition of phytoplankton in the guts of both the fishes caught during the same time from Kulgarhi reservoir was more or less the same. Zooplankton, the favourite menu of catla, formed 21 per cent of the gut contents of silver carp. The authors concluded that silver carp hampered the growth of catla in the reservoir and advocated caution before its stocking in Indian reservoirs.

It is significant to note that despite its entry into a number of Indian reservoirs, by accident or otherwise, silver carp failed to get naturalised anywhere except Gobindsagar. Considering that the reservoir, with its temperate climate, is closer to the original habitat of the fish and has a distinctly cold water hypolimnion due to the discharge from Beas, the silver carp seems to have found a congenial habitat for growth and propagation. Although introduction of silver carp was never cleared by the Committee of Experts constituted by Govt. of India, the fish is being stocked in a number of reservoirs in the country. Nowhere did the fish make an impact as it did in Gobindsagar. Therefore, fears regarding the threat of extinction of catla from the Gangetic and peninsular India posed by silver carp are perhaps misplaced.

Common Carp

The three varieties of the Prussian strain of common carp, *viz.*, the scale carp (*Cyprinus carpio communis*), the mirror carp (*C. carpio specularis*) and the leather carp (*C. carpio nudus*) were introduced in India during 1939. They were stocked in several high altitude ponds and lakes during the 1950s. Later, in the 1957, the Chinese (Bangkok) strain of the common carp was brought into the country, primarily for aquacultural purposes, considering its warm water adaptibility, easy breeding, omnivorous feeding habits, good growth and hardy nature.

Like tilapia, common carp soon found its way to all types of reservoirs in the country. Relative ease at which the fish could breed in controlled conditions prompted the departmental fish farms throughout the country to produce the seed of common carp in large numbers and to stock them in the reservoirs. However, such stocking attempts were devoid of any ecological reasoning. The Bangkok strain of common carp has been stocked in a large number of reservoirs in the plains and European strain was introduced in the reservoirs of temperate zones and high altitudes. But their performance in reservoirs is erratic, despite heavy stocking.

Common carp is not a suitable fish for stocking in Indian reservoirs, especially the larger ones, for diverse reasons. Being a sluggish fish, its chances of survival in a predator-dominated reservoir are very poor. They are not frequently caught in a passive fishing gear like gill net, due to its slow movement and bottom dwelling habit. It is no wonder, despite a regular stocking for 13 years (involving 537,000 fingerlings), not a single common carp was ever caught from Nagarjunasagar. Obviously, the stocked fishes failed to survive among the marauding predators. This has been the fate of common carp stocked in all the deep reservoirs, with a few exceptions such as Krishnarajasagar. A more important disqualification is its propensities to complete with some important indigenous carps like *Cirrhinus mrigala* and *C. cirrhosa* and *C. reba* with which common carp shares food niche. Instances where the presence of common carp has resulted in the decline of *Cirrhinus* sp. are available in Girna and Krishnarajasagar.

The mirror carp has a dubious distinction of jeopardizing the survival of a number of native fish species, after its introduction in some of the upland lakes of Kumaon Himalayas, the Dal lake in Kashmir, Gobindsagar, and the reservoirs of the north-east. In the Dal lake, common carp found a favourable environment by virtue of the shallow lake basin, extensive submerged vegetation, and rich food resources. By virtue of the specific ecological advantage, the fish propagated itself profusely to the perli of indigenous snow trouts like *Schizothoraichthys niger, S. esocinus,* and *S. curvifrons*. The snow trouts had the twin disadvantages of low fecundity and the stream breeding behaviour. Mirror carp has caused similar damage to the snow trouts in Gobindsagar reservoir and *Osteobrama belangeri* in Loktak lake of the north-east. Analogy of events related to the common carp and snow trouts sends out enough signals regarding the potential harm the former can do to the ichthyofauna in the plains.

Other Exotic Species

Three exotic carps, being considered for introduction in the country are the bighead carp, *H. nobilis* (already gained entry unofficially) the mud carp, *Cirrhinus molitorella* and the snail carp, *Mylopharyngodon piceus*, feeding on zooplankton, detritus and molluscs respectively. Natarajan (1988), after a thorough probe into the ecological implications of their introduction, considered their introduction as an irrational step, as all of them infringe on the food niche of economic carp species of India and the alien species possessed all propensities for causing extinction of their native counterparts.

Artificial Eutrophication

Fertilization of reservoirs as a means to increase water productivity through abetting plankton growth has not received much attention in India. Multiple use of the water body and the resultant conflict of interests among the various water users are the main factors that prevented the use of this management option. Surprisingly, fertilization has not been resorted to even in reservoirs which are not used for drinking water and other purposes. Documentation on fertilization of reservoirs in India is scarce. Sreenivasan and Pillai (1979) attempted to improve the plankton productivity of Vidur reservoir by the application of super phosphate with highly encouraging results. As soon as the canal sluice was closed, 500 kg super phosphate with P_2O_5 content of 16 to 20 per cent was applied in the reservoir when the waterspread was 50 ha with a mean depth of 1.67 m. As an immediate result of fertilization, phosphate content of water increased from nil to 1.8 mg l^{-1} and that of soil from 0.242 to 0.328 per cent. Similar improvements in organic carbon and Kjeldal nitrogen has been reported from soil and water phases on account of fertilization. Experiments were also conducted with urea in the same reservoir.

Application of lime was tried in some upland natural lakes for amelioration of excessive CO_2 and acidity at the bottom (Sreenivasan, 1971). This measures, together with the application of superphosphate in Yercaud lake, raised the pH of water from 6.2 to 7.3 and decreased the CO_2 in bottom water from 38 to 6.5 mg l^{-1}. There was a corresponding increase in species number and biomass of plankton.

The basic objective of fertilization is to increase the plankton density and thereby accelerating primary productivity. Fertilization in Vidur reservoir resulted in a marked increase in benthic and plankton communities and doubling of the primary production rate. After two successive applications of fertilizers, significant limnological changes took place including the presence of free carbon dioxide and decrease in pH and dissolved oxygen at the bottom layer of water. The methylorange alkalinity increased from 44 to 108 mg l^{-1} from the surface to bottom, indicating a high organic productivity. Phosphate fertilization triggered the tropholytic activities mineralising the organic matter and producing carbon dioxide. As a direct benefit from the fertilization, a 50 per cent increase in fish production, along with three-fold increase in the size (average weight) of catla, rohu, mrigal, *L. fimbriatus* and *L. calbasu* were achieved.

Experiments on fertilization is in progress in the 90 ha Naktra reservoir in Madhya Pradesh, under a research project of the Central Inland Capture Fisheries Research Institute. Organic and inorganic fertilisers are being applied to improve the water and soil quality status. Artificial eutrophication as a decisive management option was tried in India for the first time in Kyrdemkulai (80 ha) and Nongmahir (70 ha) reservoirs of the north-east (Sugunan and Yadava, 1991a, b) by applying poultry manure (10 t ha^{-1}), urea (40 kg ha^{-1}) and single superphosphate (20 kg ha^{-1}).

Fertilization can play a key role in many small reservoirs of India, which require correction of oligotrophic tendencies. A number of reservoirs in Madhya Pradesh, the north-east and the Western Ghats, receiving drainage from poor catchments show low productivity, necessitating artificial fertilization. Chinese experience in fertilizing the small reservoirs for increasing productivity has been reassuring (Yang *et al.,* 1990). In Shishantou reservoir, a management strategy comprising fertilization by organic and inorganic manures and feeding resulted in phenomenal production hike from 1 500 kg ha^{-1} to 6,000 to 7,000 kg ha^{-1} during 1985 to 1989. Before fertilization, the plankton biomass in Shishantou was 1.5 mg l^{-1}, which was raised to 6.5 mg l^{-1} through application of organic fertilizers at the rate of 6.375 t ha^{-1}. The plankton biomass, after dropping during the peak precipitation period, picked up to 20.51 mg l^{-1} during the post-rainy season months, with corresponding increase in fish production.

Pollution

Ecodegradation of reservoirs has been on the increase due to the rapid pace of industrialisation, poor environment management in the catchment and a variety of other factors. Apart from the direct entry of industrial, municipal and thermal wastes, the pollution load carried by the upstream rivers is also accumulated in the reservoirs. The environmental degradation in reservoirs is caused mainly by the waste discharge from industrial, municipal and agricultural sources and the thermal power plants (Table 18). High rate of siltation due to poor catchment management also affects the biological productivity.

Table 18: Pollution in Reservoirs

Reservoir	*Name of River*	*Sources of Pollution*
Getalsud	Subarnarekha	Heavy engineering, chemicals and sewage.
Gandhisagar	Chambal	Textile, chemicals, trade effluents from Indore, Ujjain and Kota.
Tungabhadra	Tungabhadra	Paper, iron and steel, rayon, chemicals and sewage.
G.B.Pantsagar	Rend	Thermal power plant, coal washery, chemicals.
Bhavanisagar	Bhavani	Viscose factory effluent.
Hussainsagar	Musa	Trade effluents and sewage from Hyderabad city.
Hirakud	Mahanadi	Paper mill
Byramangala	Vrishabhavati	Industrial effluents and city sewage
Sandynulla	–	Animal products

Modified from Joshi, 1990.

Thermal Pollution

A number of reservoirs have been selected, of late, as sites for thermal power plants due to their dual utility as perennial source of water supply and disposal point for heated effluents. Thermal power generation capacity of the country has been registering a steady growth of 8 per cent per annum and by the turn of the century, the installed capacity is expected to reach 84,000 MW. Various thermal plants in the country are estimated to generate 10 billion m^3 of hot water (40° to 5°C) and 17 million t of fly ash every year. Fly ash is known to contain heavy metals such as Zn (6 per cent), Ba (12.2 per cent), Cu (1.3 per cent), As (0.02 per cent), V(0.08 per cent), Ti(0.02 per cent) and Mn (0.23 per cent), which may find their way to the nearest river stretch or a reservoir.

Rihand is a large man-made lake of 46,000 ha, into which converge cooling waters from four super thermal power plants under the public sector*viz.*, Singrauli (2,000 MW), Vindhyachal (2,260 MW), Anpara (3,130 MW) and Rihand (3,000 MW), besides the private sectors Renusagar thermal power plant with a capacity of 210 MW. All these power generating plants are located within a small area of 30 km^2. Chandra *et al.* (1985) reported adverse effects of heated discharge on resident aquatic organisms. They recorded mortality of fish and decrease of aquatic life within 50 km of the discharge point, owing to high temperature (46 to 52°C) of the effluent. Deposition of fly ash has been reported up to 500 m downstream of the outfall point. Cooling waters of Renusagar power plant discharged into Rihand reservoir are acidic and high in chlorides. Although an increase in water temperature is known to cause deoxygenation, a rise within reasonable limit enhances photosynthetic activities resulting in supersaturation of water with oxygen.

The main ecological consequences of a heated water discharged into the aquatic ecosystem are increase in water temperature, change in chemical composition and change in metabolism and life history of aquatic communities. The heated discharge may elevate the water temperature by 8 to 10 °C which may cause mortality of fish and fish food organisms. Temperature also exerts direct influence on toxicity. Apart from the rise in temperature, discharged waters are often altered chemically during the cooling processes. Davies (1966) showed that cooling tower discharge has lower ammonia level, higher concentration of nitrate and TDS, and lower levels of organic nitrogen, when water is abstracted from a polluted water source.

Temperature above 40 °C has been reported to negatively affect the plankton and benthic communities. Generally, fishes avoid heated effluents discharge points by swimming away to safer places. They can also withstand wide fluctuation of temperature (8 to 10 °C). However, the reproduction of fish is affected due to deposition of fly ash in the marginal areas of the river/reservoir which act as their breeding grounds. All the power plants around Rihand reservoir are located near the intermediate and lotic sectors, where the fishes are known to congregate (Desai, 1993). The most deleterious among the impacts of thermal pollution is the blanketing effect on the reservoir bed. Fly ash covers extensive areas of the bottom, blanketing off the substratum, resulting in retardation or total elimination of benthic communities. Thick mat of fly ash deposit at the bottom bed over the years may seal the nutrients away from the water phase and thereby affect productivity.

Domestic Wastes

A number of reservoirs contiguous to towns and cities face threat from sewage pollution. Although from the fisheries point of view, organic loading within certain limits does not hamper the productivity, sewage load in excess can cause aseptic conditions, adversely affect the biotic communities, retard productivity and render the fish unfit for human consumption. Moreover, the problem needs to be addressed from public health and aesthetic points of view. The acute cases of hyper-eutrophication due to city sewage discharge in Hussainsagar, Mansarovar, Byramangala and Sandynulla reservoirs cause serious impediments in ecosystem management. Cases of heavy fish mortality is reported in Byramangala (Raghavan *et al*, 1977) and Hussainsagar (Hingorani *et al*, 1977).

The major adverse impact of sewage pollution can be assessed from deoxygenation, high BOD load, rapid eutrophication and accumulation of heavy metals in the environment. Sharp fall in dissolved oxygen in water puts the biotic communities under severe stress. While some species can tolerate a wide range of dissolved oxygen, many communities are highly sensitive to this parameter. As a chronic effect of oxygen depletion, some of the component populations are eliminated from the riverine community, causing far reaching changes in the trophic cycle. For instance, complete absence of zooplankton during January to August and its reappearance in September represented by *Keratella* sp., associated with abundance of phytoplankton like *Microcystis* sp. *Oscillatoria* sp., *Hormidium* sp., and *Nitzschia* sp., have been observed downstream of the sewage effluent outfall on the Ganga and Yamuna. The outfall area is dominated by *Chironomus*, followed by oligochaetes (*Tubifex* and *Nais*) in both the rivers, while areas below the outfall are characterised by the dominance of *Chironomus* followed by gastropods and bivalves.

Apart from affecting the organisms at lower trophic levels, intensive rate of pollution from municipal sources often causes direct fish kill, especially in small reservoirs where the problem gets aggravated due to reduced water flow rate. There are potential problems relating to the use of chlorine for disinfecting the sewage effluents for public health purposes. Owing to the increased use of synthetic detergents for domestic purposes, their incidence in the sewage effluents are on the increase. Synthetic detergents being absorbed into the body system of fish impair their growth and reproductive capacity. Detergents mixed with oil may be 60 times more toxic than the oil alone. Synergistic action of detergents with insecticides has also been recorded. Its sub-lethal concentration causes thinning and elongation of respiratory epithelial cells. Sodium lauryl sulphate is more toxic to freshwater teleosts, compared to alkyl benzene sulphonate (13-60 mg l^{-1}). Impact of heavy metals is discussed later on in this document.

Industrial Effluents

Wastes emanating from an array of industries such as chemical plants, textile mills, heavy engineering plants, paper mills, iron and steel factories, rayons, *etc.*, cause pollutional hazards in Indian reservoirs. Several instances of ecosystem damage and fish kill due to industrial effluents have been documented. Effluents from the Kanoria chemicals discharged into Rihand reservoir are alkaline (pH 9.2) and high in total alkalinity (4,770 mg l^{-1}), specific conductivity (12,816 μmhos), chlorides (5,173 mg l^{-1}) and free chlorine (1,924 mg l^{-1}) (Chandra *et al.,* 1983). These wastes have shown severe toxic effects on phyto - and zooplankton. Direct fish kills have also been reported in this reservoir due to high chlorine bearing wastes (Arora, *et al*, 1970). Effluents from a paper mill at Brajrajnagar are discharged into Hirakud reservoir and Sugar mill wastes finding their way to a small reservoir in Gorakhpur were the cause of complete replacement of carps by the uneconomic fishes (Natarajan, 1979b). Discharge of industrial wastes consisting of dissolved and insoluble solids, free chlorine and lime of Mettur chemical factory into the surplus water channel of Stanley reservoir is reported to have resulted in large-scale mortality of carps and catfishes in summer.

An industrial unit near Sandynulla reservoir manufacturing gelatin from animal bones and the effluents from this factory bearing high BOD and phosphate are discharged into the reservoir, which is already eutrophic on account of sewage wastes from the city of Ooty. The south India Viscose, manufacturing viscose rayon and staple fibre, discharges 16,000 m^3 of waste material into the river Bhavani at Sirumughai, causing pollutional hazard in Bhavanisagar reservoir, causing increase in bicarbonate and carbon dioxide and fall in dissolved oxygen and pH along with occasional fish kills. Another synthetic fibre manufacturing unit, Harihar Polyfibre discharges wastes into the tributaries of the river Tungabhadra, resulting in ecosystem degradation and fish kills (Joshi and Sukumaran, 1987). Several major and minor industries located on the river Bhadra in the industrial town of Bhadravati in Shimoga district are the source of heavy metals discharge in the inflowing water of Tungabhadra reservoir (Singit *et al.,* 1987).

Impact of Industrial Pollution

Industrial effluents, though comparatively less in volume, may cause considerable harm to the aquatic environment and the biotic communities including fish and ultimately affect man through food chain. Non-biodegradable and persistent types of pollutants like heavy metals, chlorinated hydrocarbon pesticides, oil components having high boiling points and radionuclides get more concentrated at higher trophic levels through biomagnification and pose threat to human health. Industrial effluents include a wide variety of chemical toxicants and heavy metals, apart from those contributing substantially to the BOD load such as pesticides which are used in processing the raw materials in many industries. In addition to the sub-lethal chronic effects on the environment, certain direct impacts are also discernible.

Natarajan (1979b) stressed the importance of protecting the upstream zones which are biologically sensitive areas. So are the head zones (*i.e.*, that part of the reservoir into which river flows) of the reservoirs where fish concentration is much higher. Discharge of effluents into the upstream can throw up a chemical barrier for breeding migration of economic carps, apart from causing considerable mortality to spawn and hatchlings.

Chronic Effects of Effluent Discharge

There are two general classes of effects of pollutants on water uses. Some of the dramatic effects of toxicity, including fish kills are often well-publicised. But the other class of effluxion which involves continuous chronic sublethal degradation needs a more demanding consideration. This degradation goes unnoticed except by ecologists, taxonomists and sometimes by fishermen. It is neither dramatic nor well-publicised. No gory pictures of heaps of dead fishes or large water areas covered with oil spill or debris, but the killer masquerades in the form of reduction in the rate of reproduction by aquatic species or subtle changes in the food chain pattern on which the fish populations depend. The contaminants get accumulated in the water, soil and detritus phases of the environment and get biologically magnified, as they enter into fish tissues.

The harmful industrial and municipal effluents are as diverse as they are obnoxious. There is a diversity of harmful chemical toxicants that emanate from different industrial units, the nature of substances varying, depending on the products, production processes and the raw materials used. Similarly, the agricultural runoff carry heavy load of non-biodegradable pesticides. Domestic wastes also contain a variety of chemicals, detergents and organic load. Unfortunately, the impact of these toxicants on the biotic communities is very complex and our knowledge in this regard is grossly inadequate. It is not even possible to prescribe a precise *safe* limit in respect of any of the chemical pollutants. The pre- 1960 literature in this regard was based on short-term studies and used mortality as an end point. This is no longer valid, as the emphasis has now shifted to a balanced ecosystem rather than prevention of fish kills. To prevent the ecosystem from gradual degradation, we must provide criteria that will protect the entire life cycle of the desirable species as well as the food chain on which these species depend. A significant reduction in available food or reproductive success will result in a condition similar to that after a fish kill. Criteria must, therefore, be based on chronic or life cycle studies that may also permit extrapolation to untested species or toxicants.

Where multiple discharges exist, many chemical reactions may occur in the receiving water that intensify or reduce the toxic effect of the original materials. Thus, apart from the information concerning the effluents, specific knowledge on the potential chemical and physical changes involved is imperative to estimate the effects of multiple effluents on the environment. Moreover, the potential of combined stress on aquatic life cannot be explained on the basis of a single contaminant. The problems involving pH and metal toxicity are common where toxicity increases due to decrease in pH values. Similarly, the environments barely acceptable with regard to dissolved oxygen become totally unacceptable if the temperature is permanently increased, resulting in an increase in oxygen demand by the aquatic life.

The validity of applying the existing *safe concentration limits* is rather limited, as they are generally determined under controlled laboratory conditions. In the laboratory, fish are fed *ad libitum*. They are treated prophylactically, if needed: there are no predators, no competition for spawning areas, and no exposure to extremes of natural water quality. The effect of 2, 4-D on fishes adequately illustrates the gradual unspectacular decline in the quality of aquatic life. A study conducted at the Bureau of commercial fisheries pesticide laboratory at Gulf Breeze, Florida (Brungs, 1972) indicated that fish exposed to 2, 4-D for 1 to 5 months grew and survived as well as the control animals. However, the exposure, apparently lowered the general body resistance to a microsporidian parasite and a massive invasion of the central nervous system of the fish resulted.

Pesticides and Heavy Metals

Hazardous and toxic substances such as pesticides and heavy metals are carried to the reservoirs through the effluents and the rain washings from the catchments. These substances are highly persistent and thereby contaminate the entire biogeochemical cycle of static systems like reservoirs. The problems are aggravated due to the capacity of toxic substances to get biomagnified in fish tissues, which otherwise exist in water in extremely low concentrations. Such a situation, apart from resulting in low fish, transports toxic metals and pesticides into the human body through the contaminated fish. Heavy metal accumulation in water, sediments and plant tissue has been reported from Byramangala (Table 19).

Table 19: Status of Heavy Metals in Byramangala Reservoir

Metal	*Concentration*		
	Water (μg l^{-1})	*Sediment (μg^{-1})*	*Plant (μg g^{-1} dry wt.)*
Zn	87–130	50–197	76.5–207.8
Cu	28–52	38–64	33–143
Cd	nd-15	32–106	1.4–2.1
Cr	Nd	0.88–1.32	0.42–0.7
Pb	16–22	53.4–101.2	5.3–9.0
Hg	0.08–0.12	0.14–0.4	0.29–0.63

After Joshi, 1990.

Pesticide residues have been detected in the riverine ecosystems in all the river basins of the country (Joshi, 1990). High levels of BHC, methyl parathion, endosulfan and DDT and their biomagnification in biotic communities like plankton, benthos and fish have been reported from Cauvery, Ganga and Yamuna rivers. However, such observations from reservoirs are rare. Joshi (1990) recorded significant presence of residues of different isomers of BHC, and DDT and its metabolites (DDE, DDD) in fish and plankton of Rihand reservoir. Considering that the reservoir is situated in a relatively remote place far away from agricultural and industrial activities, the observation assumes importance.

Studies conducted in Panchet reservoir, Bihar (Gopalakrishnan *et al.,* 1966) showed adverse effects of effluents from coal washings on the recruitment of Indian major carps. Sinha (1986) reported rapid eutrophication in man-made lakes due to drainage from coal fields in south Bihar.

Siltation

Excessive siltation leading to drastic decrease in the water holding capacity and even damage to concrete hydraulic structures is a common problem in reservoirs. Siltation also hampers the productivity of water body by affecting the life processes of biotic communities. Erosion of top soil in the catchment area is the main man-made factor that leads to increased sediment load in rivers. Vegetation cover on the slopes acts as an adherent of top soil during the surface runoff. Removal of forest cover through logging, grazing, road construction or for urban needs makes the soil susceptible to erosion. The entire suspended and bed load materials carried by the rivers, however, are not exclusively the contribution by man. The catchment areas, especially those of the Ganga river are characterised by a prolonged dry season followed by a turbulent monsoon, with river discharges up to 85,000 m^3 per second. Therefore, heavy erosion and high sediment load are characteristic of Indian rivers. However, the tampering of environment in catchment areas adds considerably to the sediment load and the problem needs to be addressed through appropriate conservation measures.

Suspended particles tend to settle down in the lentic waters of the reservoir causing many problems. It is estimated that in India 5,334 million t of soil is eroded every year from the cultivable land and forests. The Indian rivers carry about 2,050 million t of silt, of which nearly 480 million t is doposited in the reservoirs and 1,572 million t is washed away into the seas. Loss of storage capacity of the reservoirs due to siltation is one of the most serious consequences of soil erosion. Many of the reservoirs recorded siltation rates, much in excess of what was envisaged during the planning stage of the project, due to increased rate of sediment load in the incoming waters (Table 20).

Apart from diminishing the water holding capacity of the reservoir and cutting its life, Siltation also affects the biota by blanketing the benthic and periphytic community. It also hampers the recruitment by destroying the breeding grounds and retards the overall productivity of the ecosystem.

Cage and Pen Culture

The unconventional production systems, such as cage and pen cultures have not become very popular in India, although they have a definite role to play in augmenting fish production from open water, especially the reservoirs. It is now widely accepted that the pen enclosures erected in the reservoir margins can be used as nurseries to raise stocking material to obviate the necessity for constructing concrete nursery farms which are cost-intensive. Similarly, the rearing of fish in cages and pens up to marketable size enables easier stock manipulation and total harvesting. However, non-standardization of farm practices and the materials to be used in the operation still acts as a major retardant for large-scale adoption of these culture systems in Indian reservoirs.

Table 20: Rate of Siltation in some Selected Indian Reservoirs

Reservoir	*Rate of Silting (in ha m 100 km^{-2} yr^{-1})*	
	Assumed	*Actual*
Gobindsagar	4.29	6.00
Nizamsagar	0.29	6.57
Tungabhadra	4.29	6.11
Hirakud	2.52	3.98
Shivajisagar	3.24	15.24
Gandhisagar	3.61	10.05

After, Joshi, 1990.

Species Selection

Main criteria for the choice of candidate species for cage and pen culture are:

1. Fast growth rate,
2. Adaptability to the stresses in enclosures due to crowded conditions,
3. Ready acceptance of artificial feeds consisting mainly of cheap agricultural byproducts,
4. High feed conversion rates,
5. Resistance to diseases, and
6. Good market demand.

The candidate species should preferably not breed in the cages and upset the population balance. Under the Indian conditions, the Gangetic major carps (*C. catla, L. rohita, C. mrigala*), the chinese carps (*Hypophthalmichtys molitrix, Ctenopharyngodon idella*), common carps (*Cyprinus carpio*), the magur (*Clarias batrachus*) and tilapias satisfy these requirements to a great extent. Murrels (*Channa* spp.) also can be cultured in maritime States, where marine trash fish is available at a discount. Selection of species, however, is mainly dictated by the local demands and availability of quality seed and other inputs in adequate quantities.

Site Selection

Appropriate site selection is important for successful enclosure aquaculture.

Sheltered, weed-free, shallow bays are the ideal locations for installing pens and cages. The sites should have adequate circulation of water, with wind and wave action within moderate limits. Excessive turbulence may lead to wastage of fish energy for stabilizing themselves and loss of feed. The other major considerations are that the water should be pollution-free, availability of seed in the vicinity, easy accessibility to the site and a ready market for fish. Flowing waters with a slow current of 1.0 to 9.0 m per minute are considered ideal for cage siting. It is desirable to install cages a little away from the shore to prevent poaching and crab menace.

Water level fluctuation is the most important consideration in site selection for the pen culture operations in reservoirs. A scrutiny of the contour map and the monthly fluctuation patterns of reservoir levels will enable the location of suitable sites, which retain sufficient water for the required period of time. Sites which dry out during summer will be ideal, as it is easier to erect pens on dry land, to be inundated later as the water level increases. Similarly, some bays of the reservoir retaining water for sufficient period can be identified and cordoned off by erecting barricades.

Cage Culture

Experiments on cage culture conducted in India have been exploratory in nature and the yields obtained, so far, are not impressive. The supplemental feeds given are oilcakes, ricebran, soy bean flour and silkworm pupae, which have great demand in cattle, poultry, pig rearing and other animal husbandry practices and hence command a good price in the market. The food quotient obtained in the cage culture of various species has not been high, except in the case of tilapia, making conventional supplemental feeding unremunerative. The low production and feed conversion rates are mainly due to the relatively low stocking density and many deficiencies in the feed. The feed is often not in a water-stable form and nutritionally balanced to promote growth. There is need for evolving suitable complete feeds for individual species of fish from the locally available raw materials, by experimentation.

One of the major constraints of the cage culture system is the lack of suitable cage designs to withstand severe wave action, common in Indian reservoirs. Mukherjee (1990) suggested a number of flexible, floating barriers, sheet barriers and rigid floats to protect the cage structures from wave action. The floats dampen the wave thrust and absorb the wave energy before the wave can propagate and strike the cage and cause damage. Kumaraiah and Parameswaran (1985) proposed a circular cage that could be used in reservoir with moderate wave action for culture of carps, tilapia and air breathing fishes. The cages can float at the surface, remain just submerged or rest at the bottom. Floating cages are considered to be most appropriate for Indian conditions and all the experiments conducted so far in the country for seed rearing, growout, nutrition and biomonitoring have been in such enclosures.

Cage Materials

Floating fish cages can be constructed out of a variety of materials including metal, wood, bamboo and netting. Fairly fine-meshed nylon netting is used for nursery purposes. Cages made of monofilament woven material of 1.0 to 3.0 mm mesh size are light and easy to handle but last only for six months to one year, depending on their thickness. Knotless nylon webbing of 3 to 6 mm mesh size and knotted nylon webbing of 7 to 15 mm mesh have been found to be very durable as cage material. A battery of cages can be buoyed up within a bamboo catwalk which will serve as a working platform, floated by sealed empty barrels. Circular and boxlike cages of varying dimensions on conduit pipe structures which can be easily assembled, and suitable floatation systems have been designed in India. Similarly, self–floating cage with HDPP pipe structure has also been experimented with succesfully.

In Jari tank near Allahabad, nylon cages (20 mesh cm^{-1}; size 2.2×1.6×1.45 m) were stocked at a density of 8500 hatchlings m^{-2} (size 6.5 to 7.8 mm). These grew in 21 to 28 days to 30.2 to 45.6 mm with a survival of about 25 per cent (Anon., 1979). In fry rearing, the stocking rate in the cages (mesh size 3 mm) was 700 to 2 500 m^{-2} and within 90 days they attained a size of 103.6 to 121.8 mm. The feed given was powdered soybean, groundnut cake and rice bran in equal proportions. Rearing of carp fry was done in Getalsud reservoir, where they (10 to 31 mm in size) were stocked in 2.4 × 1.5 × 1.5 m cages at the rate of 300 to 700 m^{-2}. The growth rate per month was 17, 25 and 20 mm in mrigal, catla and rohu respectively. The stock was fed with mustard and groundnut cake and rice bran in the ratio 3:1:1 at 30 per cent of the body weight (bw) of the stock for 4 days and thereafter at 20 per cent for the rest of the period. Summary of cage culture experiments conducted is presented in Table 21.

A series of cage culture trials have been reported from a 12 ha impoundment in Bangalore (Parameswaran, 1993). In an experiment conducted with monofilament cloth cages of size 10.5 m^{-2}, common carp and silver carp fry were reared at a ratio of 40:1 at a stocking density of 225 m^{-2}. Put ona diet of powered rice bran, defatted silkworm pupae, groundnut cake and soya flour in 12:5:2:1 ratio at 10 to 20 per cent bw day^{-1}, the survival obtained at the end of 4 months rearing was 97.5 per cent in common carp and 88 per cent in silver carp. The stock attained average final weight of 20 and 8.6 g respectively. However, experiments conducted on catla gave erratic results with survival rate varying from 9 to 71.4 per cent. In another trial, 30,000 spawn obtained from cage grown common carp parents were reared in 4.5 m^3 auto-floating (PVC frame) monofilament cloth (mesh: 15 cm^{-1}) cages. In 35 days, the fry attained a size of 25.4 mm with 38 per cent survival. Restocked in 3.5 m^3, 8 mm mesh knotless nylon netting cages, at a density of 475 m^{-3}, they grew to 54.8 mm/4.9 in 75 days with a survival rate of 88.5 per cent.

In a cage culture experiment reported from Tamil Nadu (Parameswaran, 1993), 10 days old fry (size 10 mm) stocked at a density of 500 m^{-2} were raised to a size of 50 to 60 mm in 40 days, with survival rates ranging from 45 to 85 per cent.

Table 21: Summary of Growout Experiments Conducted in Cages in India

Species Cultured	Cage Vol. (m^3)	Stocking Density (m^{-2})	Mean Harvest Size (g)		Culture Period (Months)	Production (kg m^{-2} $month^{-1}$)	Feed	Feeding Rate (per cent bw)	FCR	Reference
Cyprinus carpio	15.75	30–38	40–50	325	6	1.55–2.22S	WP, GNC, RB (8:9:3)	10–20	8.3–10.4	Govind (MS.) 1983
Catla catla	15.0–15.75	13–49	8–50	544–772	6–8	0.83–1.30	GNC, RB (1:1)	5–10	5.6–6.6	Govind *et al.*, 1988
Hypophthalmichthysm olitrix	10	15	61	472	10	0.7	SWP, RB, GNC (1:2:3)	3–5	3.1	Kumaraiah *et al.*, 1991
Labeo calbasu	10	5	16.5	208	8	0.1	GNC, RB (1:1)	2	2.9	Kumaraiah *et al.* (unpublished)
Ctenopharyngodon idella	3	33–67	7–10	350–400	6	2.0–3.3	Lemna, Hydrilla	80	–	Bandopadhyay *et al.*, 1991
Oreochromism ossambicus	5–10	100–200	6.0–7.6	32–62	2–5	09–1.6	RB, GNC, CFP (1:1:1)	3–5	1.8–2.3	Kumaraiah *et al.*, 1986
Channa marulius	5	40	25.8	177	5.3	0.8	Trash fish	10–12	2.5	Kumaraiah Parameswaran; and (unpublished)
Clarias batrachus	2	100	7.4	36.9	3	1	–	–	–	Murugesan and Kumaraiah, 1972

SWP: Silkoworm pupae; GNC: Groundnut cake; RB: Rice bran; CFP: Cattle feed pellets; FCR: Food Conversation Ratio.

Department of Fisheries in Tamil Nadu has been undertaking rearing of spawn and fry of major carps in floating cages during July to September every year. However, data on the stocking density, nutrition, growth and survival are not available.

Rearing of the fry of Indian major carps was tried in Tungabhadra reservoirs in the year 1984–85 (Singit *et al.,* 1985). Four floating cages, made of 16-P velon screen fitted on rectangular bamboo frame of 10×4×1 m, were stocked with rohu and mrigal. Although survival rates ranging from 37.5 to 87.5 per cent were obtained at the end of the 3 months rearing period, the experiment was vitiated due to the destruction of cages due to heavy winds. At stocking densities ranging from 2 to 5 million ha^{-1}, growth of about 100 mm (33 g) was obtained.

Dependent on the type of management input, fish production rates obtained for growout in cages vary greatly. Unlike the hi-tech system of saturated stocking and feeding on enriched formulated diets, the production recorded in cage culture of common carp is 35, 37.5 and 25 kg m^{-3} $month^{-1}$ respectively in Japan, Germany and the Netherlands. In Asia, in general, only semi-intensive and low cost technologies are adopted, mainly due to economic considerations. In India, the growing season is almost year round, except for December–January in northern parts, where the temperature is low during these winter months.

Pen Culture

Pen culture has a special relevance in reservoir management, since it has been widely recognised as a means to rear, *in situ*, the fingerlings for stocking. The number of fingerlings required for stocking the reservoirs in the country is so enormous that it is impossible to raise all of them in land-based nursery farms which makes pen nurseries *sine qua non* for reservoir management. Nevertheless, pen culture on a regular basis has not been practised anywhere in India except at Tungabhadra reservoir. The factors that hamper the standardisation of pen culture technique are:

1. The steep level fluctuations,
2. Wind and wave action,
3. Lack of suitable pen materials,
4. Weed infestation and the related harvesting problems, and
5. Non synchronisation of suitable water levels and the spawn availability.

The water retention time is important, since the rearing has to be completed before the water level in the pen goes down the critical limit. In reservoirs with high drawdown, the water retention time is very limited. Sometimes the filling takes place so late that no spawn of desirable carps will be available when the water level attains the desirable limit. The pen walls limiting the water circulation to some extent, the accumulated feed and fertilizers case eutrophication leading to weed infestation fouling of water and fish kills.

Pen Culture in Tungabhadra Reservoir

Despite all the limitations, pen nurseries are used with remarkable success in

Tungabhadra reservoir for the last 12 years. During 1992–93, 21 pens were erected in Ladakanabhavi, 25 km away from the dam site, covering a total enclosure of 3.3 ha. The pen site is situated at an elevation of 496 m above MSL and the installation was completed in the month of July, when the site was still exposed. Later, when the water level increased, the pen got inundated.

The pen area was pre-treated with organic manure that resulted in a rich growth of plankton after the filling. A total of 15 million spawn were stocked in the pens, comprising 6.75 million *Labeo rohita*, and 8.25 million *Cirrhinus mrigala*. After a rearing period of 90 days, 2. 41 million fingerlings were collected from the pen and released into the reservoir. This included 1.085 million *L. rohita* and 1.325 million *C. mrigala*, worth Rs. 495 875. Pen culture operations on similar lines are being in Kyrdemkulai and Nongmahir reservoirs of the north-east.

Seed rearing experiments were conducted in a split bamboo pen enclosure of 247.5 m^2 reinforced with a nylon netting in Punjar swamp, adjoining the Bhavanisagar reservoir (Abraham, 1980a). The pen was stocked with the spawn of *C. mrigala* (size 7 mm and *L. fimbriatus* (size 5 mm) at the rate of 4.6 million ha^{-1} and usual farm practices were followed. In 30 days, mrigal attained a size of 38 mm and *L. fimbriatus*, 28 mm. At the time of conclusion of the study after 3 months, the former had attained a size of 88 mm and the latter, 75 mm. The overall survival obtained was 27.8 per cent.

A pen culture experiment for raising catla and rohu in Manika maun, a floodplan lake in Gandak basin yielded a (computed) production of 4 t fish ha^{-1} in six months. The experiment was conducted in a bamboo screen pen (1000 m^2) and the stock was fed with mixture of rice bran and mustard cake, apart from a feed formulated from the aquatic the weeds collected from the lake.

Crafts and Gear

Gear

The presence of underwater obstacles restricts the use of active gear in reservoirs and the choice is often limited to passive gear such as simple gill nets. The most common among them is the *Rangoon* net, an entangling type of gill net without a foot rope. Another entangling type of net used in reservoirs is *uduvalai*, which has a reduced fishing height and is usually operated in shallow marginal areas to catch small fish. Shore seines of various dimensions and mesh sizes are employed in many reservoirs.

Although a number of other fishing gear such as long lines, hand lines, pole and line, cast nets, dip nets, *etc.* are in use, their contribution to the total catch is very insignificant.

Unlike the marine fisheries, very little attention has been paid over the years towards improvement of gear in the inland sector, barring an attempt to upgrade the reservoir fishing gear by two experts under the aegis of FAO/UNDP programme during the 1960s (Gulbadamov, 1962). A number of improvements have been suggested by the experts to the fishing techniques followed by the reservoir fishermen. Apart from the introduction of frame net, they have suggested improvements on the design of gill nets, beach seines and long lines, Unconventional methods such as electrical fishing, use of light as fish lure, and the use of echosounder for fish detection and survey of bottom topography were also suggested.

The main emphasis of gear improvement was the modification of gill nets. Gulbadamov (1962) designed two nets *viz.*, Sebgul I and Sebgul II, which were gill nets with modified rigging patttern. While ensuring a proper fixing of webbing on head and foot ropes, a uniform hanging coefficient was ensured. Similarly, the sideways movement of the webbing was checked to maintain effective area of the net. Ranganathan and Venkataswamy (1967) conducted experiments in Bhavanisagar reservoir to check the efficacy of the new design and found no appreciable superiority for Sebgul nets over the *Rangoon* nets.

The Central Institute of Fisheries Technology has experimented with gill nets of various colours and found yellow and orange coloured nets yielding better catch than the white coloured ones (Table 22). It has been observed that 77 per cent of the carps were caught by entangling and the rest by gilling. The usual method of increasing the entangling capacity of gill net is by decreasing the slackness of webbing, which can be achieved by suitable modifications in the hanging of the net (Nayar, 1979).

Table 22: Fish Catch by Gill Nets of different Colours

Colour of Net	*Catch kg 1000 m^{-2}*
White	38.9
Blue	55.1
Green	66.9
Orange	83.4
Yellow	85.5

After Nayar, 1979.

Alivi, the giant drag net of Tungabhadra reservoir is described in the chapter on Karnataka. This giant shore seine catch fishes of all hues in large numbers, including the juveniles of commercially important carp species. Similar nets are used in Rihand and Keetham reservoirs, where they are known as *mahajal*.

Trawling

Trawling has been attempted only in two reservoirs *viz.*, Gandhisagar and Hirakud. The findings of Kartha and Rao (1990) with regard to species selectivity of different type of trawling are interesting. The findings suggest efficacy of various types of trawling in increasing the productivity of commercial carps and checking the predator and weed fish populations. After experimental trawling with *single-boat bottom trawling, two-boat bottom trawling* and *two-boat mid-water trawling* under different speeds, it has been found that more than 92 per cent of the total catch consisted of economic varieties such as catla, rohu, murrels, mullets and featherbacks in two-boat mid-water trawling. This was in sharp contrast to the bottom trawling, of both single boat and two boat variety, which yielded mostly (64 to 91 per cent) the noncommercial species of fish. The two-boat mid-water trawling at a speed of 3 to 4 knots have been recommended for exploitation of commercial species and single and two-boat bottom trawling at 2 to 3 knots for eradication of uneconomic species of fishes.

Fishing Crafts

Coracle, a saucer shaped country craft, is the major fishing craft used in the reservoirs of peninsular India. It is made of a split bamboo frame, covered with buffalo hide. Apart from being simple and inexpensive, coracle is durable and has very good manoeuvrability in choppy waters. It is also a versatile craft used for laying and lifting of nets, besides navigation and transport of fish and other material. Coracles of Krishnarajasagar are prepared by wrapping HDPP over the bamboo frame with the help of coal tar as an external covering in place of hide. This modified version of coracle is cheaper and more durable (Anon., 1984b; Parameswaran and Murugesan, 1984).

Unlike Gobindsagar, where all the fishermen possess their boats, reservoir fishermen, in general, are too poor to own boats. In many reservoirs like Vallabhsagar and Hirakud, the fishermen could get their boats with the help of subsidy and other financial assistance from the Government or fundings agencies. In Vallabhsagar, boats are distributed by the State among the fishermen at a subsidy of 50 per cent, while in Hirakud, they get it from various schemes under NABARD and NCDC. Wooden boats are used for fishing in a number of reservoirs, especially in the North India. Flat bottomed, locally fabricated boats ranging in length from 3 to 7m are used in Kyrdemkulai, Hirakud, Malampuzha, Gobindsagar, Gandhisagar and Rihand. A plank-built, flat bottomed canoe, 2 to 3 m in length is the most popular fishing craft of Gandhisagar. In the same reservoir the repatriates from the erstwhile East Pakistan used the Bengal type dinghy, which is 5 to 7 m in length and have the additional facility of setting sails for wind propulsion.

Mechanised boats are not used in reservoir fishing in any appreciable extent. A 9.1 m long wooden, mechanised boat has been introduced by the CIFT in Hirakud reservoir, but they are too expensive for the fishermen. It is significant to note that large water bodies like Nagarjunasagar, Tungabhadra and Krishnarajasagar have no motorised craft neither for fishing nor for fish transport.

Dugout canoes, carved out of palm trees are used in Yerrakalava reservoir. In most of the reservoirs in the country the fishermen rely on improvised materials. Reservoir fishermen show considerable ingenuinity in fabricating makeshift rafts out of discarded old tyres, logs, used cans *etc.* In a vast majority of Indian reservoirs, where the catch is not very remunerative, no boats are used and the fishermen depend entirely on these improvised devices.

Other Management Measures

Pre-impoundment Surveys

Until recently, pre-impoundment surveys conducted in India in connection with dam construction invariably lacked a fisheries perspective. Faunistic surveys of the river stretches carried out before dam construction in Tungabhadra (Chacko and Kuriyan, 1948), Bhavanisagar (Chacko and Dinamani, 1949), Pipri and Rihand (Hora, 1949), Damodar Valley Corporation reservoirs (Job*et al.*,1952), Hirakud (Job *et al.,* 1955) and Gandhisagar (Debey and Mehra, 1959) were not comprehensive and did not help in any way the conservation and developmental efforts. A complete survey comprising the fish and fisheries, together with inventories of fishing villages and fishermen populations, and fishing craft gear for planned development is lacking in most cases.

The pre-impoundment surveys provide the framework for future development policies and should encompass:

1. The native ichthyofauna in the river stretch above and below the dam, and their likely chances of survival,
2. Breeding habits of fishes and the possible impact of impoundment on their recruitment.
3. Survey of breeding grounds in relation to submergence, both above and below the dam,
4. Hydrobiological characteristics of water and soil with special emphasis on the nutrient and thermal regime,
5. Needs for creating infrastructure such as, hatcheries, nurseries, ice plants *etc.*,
6. Site selection for pen nurseries, cages *etc.*, and
7. Possibilities for cleaning the area of submergence of trees and other obstructions.

Holistic pre-impoundment survey for fisheries development is a new concept in India. A beginning in this direction has been made by the Narmada Control Authority. A recently concluded socio-economic survey of the Narmada basin tried to address the problem of fisheries development with a holistic approach.

Timber Clearance

Opinion is divided on the wisdom of removing timber from the reservoir bed. While it is mostly appreciated that a reservoir bed free from obstructions facilitates the use of active fishing gear and leaves room for many other management options, many workers feel the necessity to leave at least the non-commercial timber intact for a variety of purposes such as, reducing wave action, flocculating the colloidal clay turbidity, providing habitat for fishes and substrata for periphyton deposition (Bhukaswan, 1980). Timber clearance has been tried in a number of reservoirs in

India, both before and after the impoundment. In Chillar and Benisagar reservoirs of Madhya Pradesh, trees were cut from the lake bed and auctioned before the reservoirs were filled. Harsi, Jamoia and Ghatera reservoirs are examples of complete clearance of date palm trees from the marginal areas during the summer months. Forest area of about 61. 4 km^2 were cleared in Hirakud, when the bed was exposed during drawdown.

Exploitation Systems

Fisheries being a state subject, management of reservoir fisheries vests with the State Governments. There is a great deal of divergence in the management practices followed by individual States which vary from outright auctioning to almost free-fishing. Cooperative societies (primary and apex) and the State level Fisheries Development Corporations are also involved in the fishing and marketing operations. Involvement of the above agencies and their role in fishery operations and market interventions often vary from one reservoir to another within the same State. Some sort of uniformity in fishery regulations among various categories of reservoirs as well as the need to monitor the socio-economic aspects of reservoirs fisheries.

Commercial exploitation systems followed in different States can be broadly classified under four headings *viz.*, 1) departmental fishing,2) lease by auctioning, 3) issue of licences to cooperative societies or individuals and 4) fishing on a royalty basis (crop sharing). Direct departmental fishing being not an economic proposition, is followed only in a very few reservoirs. In some reservoirs like Hirakud, Nagarjunasagar, and DVC, this practice is partially resorted to for experimental and exploratory purposes. In most of the cases, the Department exerts its control over the exploitation by acting as a marketing link and controlling the fishing effort. In Rajasthan, Madhya Pradesh and Uttar Pradesh, the small reservoirs are mostly auctioned on an yearly basis. In a number of large reservoirs, free licences are issued to fishermen without any limit. This virtual *free for all* system has been found to be detrimental to the interests of the ecosystem and fishermen in Nagarjunasagar, Yerrakalava and a large number of other reservoris in Andhra Pradesh. Crop sharing is a very popular mode of exploitation in Tamil Nadu, where, the fishermen are provided with all fishing implements, in return of which they pay a royalty (sometimes up to 50 per cent of the catch) to the Government.

Input-Output Relationship

Evaluation of the socio-economics of reservoir fisheries is a very tedious task due to the multiplicity of agencies involved in reservoir management. Reservoir fisheries is developed basically on capture fisheries lines, following the common property norm. Like the rivers, lakes and the seas, the biological wealth is considered as a nature's endowment and the State's intervention in developmental activities benefits the poor fishermen who toil in water. The investment made in developing reservoir fisheries shall be viewed in the light of the social benefits it accrues in the form of:

1. Rehabilitating the displaced population,
2. Improving the living conditions of fishermen, and
3. Providing employment opportunities.

Capture fisheries activity of the reservoirs is akin to the extractive industries like coal, oil, iron ore *etc.*, where the yield depends on the state of technology involved and the quantum of labour and capital deployed. But the renewable nature of the resource and the intricate biological principles involved in the ecosystem management imparts a heavy element of challenge into the reservoir fisheries management. Human intervention being less intense in reservoirs, compared to the aquaculture operations, yields often display violent fluctuations, even if the effort in terms of labour and capital is kept constant. Therefore, production-function relationship is bound to be intricate and less precise. A certain measure of stability needs to be imparted in production by affecting sustained improvement in yield along with remunerative returns to fishermen by narrowing the price spread between the producer and the consumer.

In aquaculture, it is estimated that 77.23 per cent of the price paid by the consumer is received by the producer (Paul, 1990). As opposed to this, a major chunk of the price is siphoned away by the wholesalers and other market intermediaries in reservoir fisheries. A study of seven reservoirs (Paul and Sugunan, 1990) for a period of six years has brought to light the major factors that determine the remunerativeness of fishing in reservoirs. Reservoir fisheries is a sector, where the chief input is labour, besides a marginal depreciation of crafts and gear. Even if the costs of stocking and other developmental measures are taken into account, this area does not call for heavy investment as in the case of pond culture.

A factor that can bring serious distortions in the income level of fishermen is the over–concentration of fishermen in the wake of low fish productivity. For instance, in Ukai reservoir, with an area of 36,525 ha, 306 boats with 3,400 gill nets (50 m each) were operated during 1985–86 to 1982–83. In a fishing year comprising 260 days, 1836 fishermen netted out 174 t of fishes. In sharp contrast to this, 520 fishermen of Nagarjunasagar shared a catch of 170 t in a year. In lower Aliyar, 17 t of fishes were harvested by 14 fishermen, each of them, after meeting the royalty obligations, could take home only Rs. 1,000 to 1,400 a year. A better picture, however, emerged in Bhavanisagar, where 80 fishermen after sharing 150 to 300 t of fishes, earned an annual individual income of Rs. 8,175/(Paul and Sugunan, 1983).

There is a need to dovetail the twin objectives of conservation and yield optimisation in reservoir fisheries management. While the fishermen and the fish merchants strive to increase the production for economic considerations, it is the responsibility of the State to ensure that economic expediency of development does not mar the ecological reasoning. Virtual free fishing, as followed in Andhra Pradesh is counter–productive to the norms of conservation and yield optimisation. Although there are fair possibilities of linking reservoir fisheries development

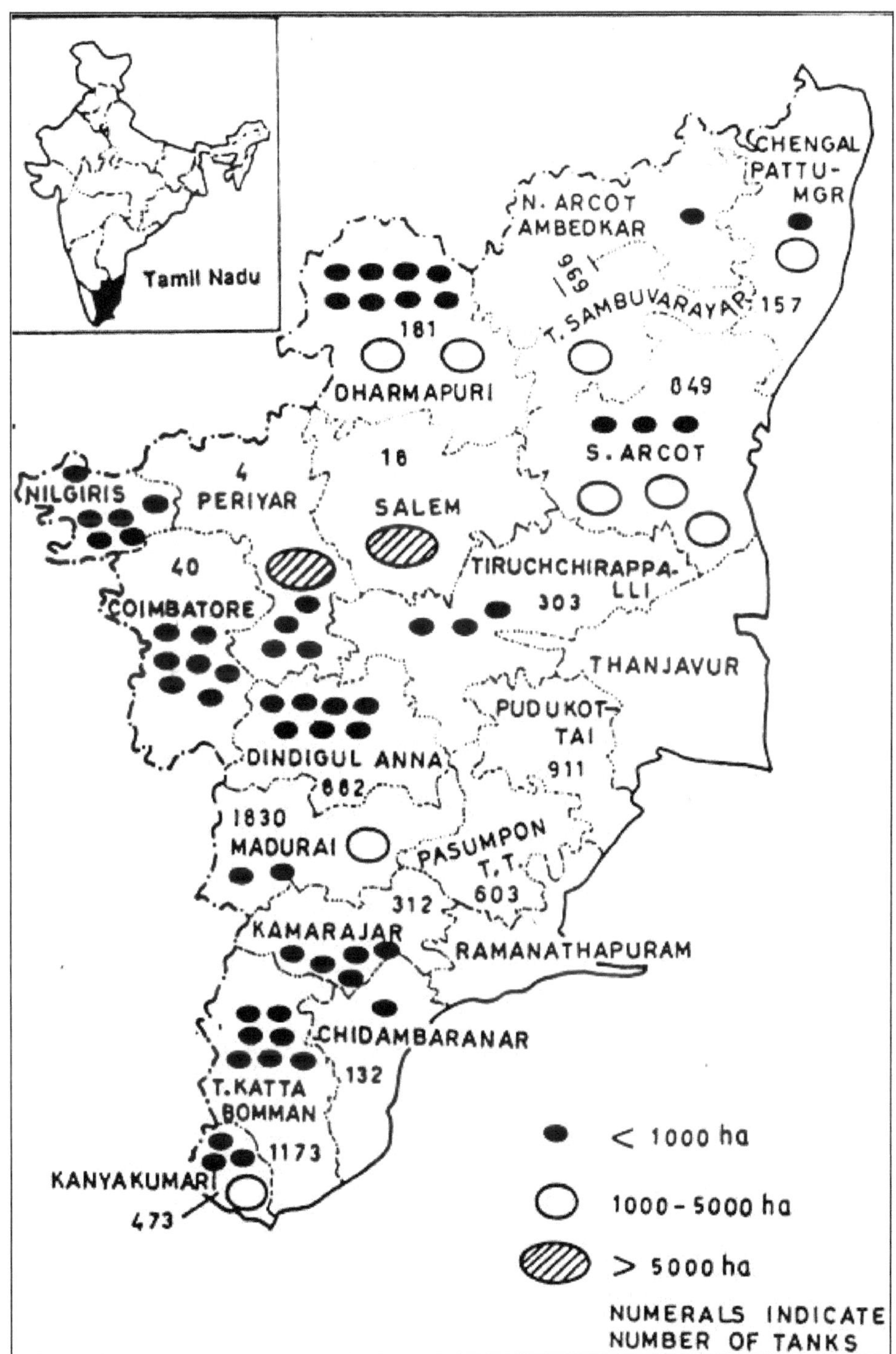

Figure 13: Distribution of Reservoir and Tanks in Tamil Nadu.

with poverty alleviation programmes, the progress made so far in this direction is not very encouraging. Chances of creating additional employment are not much in majority of Indian reservoirs. On the contrary, many reservoirs have surplus manpower which can be diverted to others which can absorb more men without eroding the income level of the existing fishermen (Paul and Sugunan, 1990).

Chapter 5

Fisheries of Important Reservoirs

5.1. Tamil Nadu

The south-eastern State of Tamil Nadu is an important geo-political, economic and cultural constituent of India. From the reservoir fisheries point of view, the State occupies a pre-eminent position, in as much as it has an ancient tradition of creating impountments for irrigation. Tamil Nadu is also in the forefront of research and development of reservoir fisheries in the country. Some of the pioneering research studies on limnology and fisheries of reservoirs were conducted in the State.

Situated in the rain shadow behind the Western Ghats, most part of the State is semi–arid and dry, drained mostly by the east–flowing seasonal, rain-fed streams. North-east monsoon is more important to the State than the southwest, thus making Tamil Nadu unique among the Indian States. The districts of Chengalpattu MGR, South Arcot, Thanjavur, Madurai, Ramanathapuram and Tirunelveli Kattabomman receive rainfall mainly from the north-east monsoon, while North Arcot, Salem, Coimbatore and Tiruchirapally in the central region depend on both monsoons. North Arcot and Thanjavur districts receive rich rainfall of 102 to 114 cm per annum, while the range of annual rainfall in the rest of the State is 640 to 890 mm. Despite the limitations in water resources, the state has made great strides in agriculture. Tamil Nadu's rice yield of 2.5 t ha^{-1} is among the highest in India and its sugar cane yield of 100 t ha^{-1} is a world record.

People of Tamil Nadu learned to conserve water and store it for irrigation since time immemorial. The landscape of the State, especially in Madurai, North and South Arcot, Dindigul Anna, Tirunelveli Kattabomman and Pudukottai districts is dotted with thousands of small impoundments (locally called *tanks*) created to store the surface flow for irrigation. Ganapati (1956a), quoting Madeley (1914) reported as many as 141 *minor irrigation tanks* or reservoirs in the 363 km^2 catchment of Red Hill reservoir. Madurai and Ramnad districts of Tamil Nadu have the maximum number of tanks and these areas are called the *land of irrigation tanks* (Pandian 1987). The Cauvery river was tamed in the first century A.D., when the Grand Anicut was built of stone and mud, covered with an outer facing of dressed granite, set in lime mortar. Upper and Lower Anicuts, the two barrages on the Cauvery downstream, were constructed by Sir Arthur Cotton in 1836 and the Mettur dam came into being in the 1930s. Even in the recent past, a large number of concrete masonary and earthen dams were built in the State, both before and after independence, resulting in the creation of 58 small and 11 medium and large reservoirs.

Table 22: Number and Area of Reservoirs in Tamil Nadu (by category)

Category	*No. of Units*	*Area (ha)*	
		At FRL	*Average*
Small reservoirs			
<500 ha	46	6,858	3,594
500– 1,000 ha	58	15,663	8,628
Total	**58**	**22,521**	**12,222**
Medium reservoirs			
1,000–5,000 ha			
Total	**9**	**19,577**	**6,374**
Large reservoirs			
5,000–10,000 ha	1	7,876	3,208
>10,000 ha	1	15,346	9,324
Total	**2**	**23,222**	**12,532**
Grand total	**69**	**58,462**	**27,534**

Fisheries Department of Tamil Nadu, with one of the best resource data bases in the country, has surveyed almost all the man– made lakes in the State, most of them being small irrigation reservoirs. Fifty-eight, out of sixty–nine reservoirs in the State fall under the category of small reservoirs (<1000 ha). Even among them, the reservoirs less than 500 ha in size dominate. However, the small reservoirs form only 27 per cent of the total water surface area. One reservoir in the category of 5000-10,000 ha (Bhavanisagar) and one in the >10000 category (Mettur) form 40 per cent of the total water surface area. There are 9 reservoirs in the medium category (1000 to 5000 ha), covering an area of 19 577 ha Most of the earlier accounts on reservoir fisheries of the State seem to have overlooked the water bodies designated as *tanks*. In the southern States of Karnataka, Tamil Nadu and Andhra Pradesh, many small irrigation reservoirs in the country side, created by erecting small earthern dams to obstruct stream flow, are generally referred to as tanks. Many of them have a very small catchment, restricted to the immediate surroundings and the others receive water from small streams and rivulets. Tanks also include some community ponds of the villages and those attached to the temples. Nevertheless, vast majority of them are man-made impoundments created by obstructing surface flow, falling under the broad definition of reservoirs. Although some lakes of tectonic origin, meteorite craters, *etc.*, must be masquerading among them, it is difficult, for the fisheries officials of the lower and middle rank, to tell the actual reservoirs from other lentic formations.

Small Irrigation Reservoirs (Tanks) (Classification on the basis of level at FRL)

The Fisheries Department of Tamil Nadu has a village-wise inventory of all tanks in the State, and that is categorised into *short seasonal* and *long seasonal* (*major irrigation*) tanks. The short seasonal tanks having an average size less than 10 ha, retain water only for a period less than 9 months in an year, while the long seasonal ones have an average size of 34 ha and retain water for 9 to 12 months. In some of the districts, the major irrigation tanks have an average size of 156 to 222 ha. Total waterspread under tanks in 22 districts of the State is estimated at 740 182 ha, of which 439 904 ha are short seasonal, with the size of individual units less than 10 ha in most cases. These small tanks have been tagged with the ponds. There are 8 837 long seasonal (major irrigation) tanks, covering 300 278 ha, situated in 17 districts, which are *de facto* reservoirs and so treated in this work. Madurai, Tirunelveli Kattabomman, Pudukottai and South Arcot districts have maximum number of perennial tank while the average size of tanks in Chengalpattu MGR, Salem and Pasumpon districts is higher.

5.2 Distribution of Reservoirs by Districts

Reservoirs of Tamil Nadu are scattered in 16 districts. The two large reservoirs (> 5,000 ha) *viz.*, Stanley (Mettur Dam) and Bhavanisagar are situated in Salem and Periar districts respectively. Dharmapuri has the maximum number of reservoirs comprising 8 small and 2 medium reservoirs, with a total area of 3 600 ha; followed by Coimbatore, Dindigul Anna and Tirunelveli Kattabomman districts with 7 reservoirs each. South Arcot district with just 3 each of small and medium reservoirs has a waterspread of 9 933 ha (Table 24). While taking irrigation tanks also into consideration, Pasampon Thevar Thirumagan District has the maximum water area (48 734 ha), closely followed by Pudukottai and Madurai districts with 46 278 and 41 508 ha respectively. The State has 8 906 water bodies in total, spread over 358 740 ha at FRL (Table 25). Eleven medium and large reservoirs in 8 districts and 58 small reservoirs in 14 districts along with their annual fish production and major fisheries are listed in Tables 26 and 27 respectively.

Table 23: Distribution of Major Irrigation Tanks in Tamil Nadu

Sl.No.	District	No.	Area (ha)	Average of a Unit
1.	Chengalpattu MGR	457	34 931	222
2.	N.Arcot Ambedkar	969	20 328	21
3.	T. Sambuvarayar			
4.	Dharmpuri	181	13 960	77
5.	South Arcot	849	17 695	21
6.	Salem	18	2 805	156
7.	Thiruchirapalli	303	10,063	33
8.	Periyar	4	275	69
9.	Coimbatore	40	1 792	45
10.	Dindigul Anna	882	12 915	15
11.	Madurai	1 830	41 508	23
12.	Kamarajar	312	17 328	55
13.	T.Kattabomman	1 173	26 683	23
14.	V.O.Chidambaram	132	1 990	15
15.	Kanyakumari	473	2 993	6
16.	Pasumpon	603	48 734	81
17.	Pudukottai	911	46 278	51
	Total	**8 837**	**300 278**	

Table 24: Distribution of Reservoirs in Tamil Nadu Districts

District	No. of Reservoirs		Total Area (ha)	
	Small	Medium and Large (L)	At FRL	At. Av. Level
Chengalpattu MGR	1	1	4163	1931
North Arcot Ambedkar	1	0	678	339
T. Sambuvarayar	0	1	2010	56
Dharmpapuri	0	1	2010	56
South Arcot	8	2	3600	1444
Salem	0	1(L)	15346	9324
Thiruchirapally	3	0	227	102
Periyar	4	1(L)	8786	3766
Coimbatore	7	0	3 72	1768
Nilgiris	6	0	486	253
Dindigul Anna	7	0	829	546
Madurai	2	1	2736	1768
Kamarajar	5	0	933	400
Tirunelveli Kattabomman	7	0	1722	883
V.O. Chidambarar	1	0	657	328
Kanyakumari	3	1	3184	1649
Total	**58**	**11**	**58 462**	**27 534**

Table 25: Number and Area of Reservoirs and Perennial Irrigation Tanks in Tamil Nadu Districts

Districts	*Reservoirs*		*Irrigation Tanks*		*Total*	
	No.	*Area (ha) (FRL)*	*No.*	*Area (ha)*	*No.*	*Area (ha)*
Chengalpattu MGR	2	4 163	157	34 931	159	39,094
North Arcot Ambedkar	1	678	969	20 328	970	21,006
Dharmapuri	10	3 600	181	13 960	191	17 560
South Arcot	6	9 933	849	17 695	855	27 628
Salem	1	15 346	18	2 805	19	18 151
Thiruchirappally	3	227	303	10,063	306	10 290
Periyar	5	8 786	4	275	9	9061
Coimbatore	7	3 172	40	1 792	47	4 964
Nilgiris	6	486	–	–	6	486
Dindigul Anna	7	829	882	12 915	889	13 744
Madurai	3	2 736	1830	41 508	1 833	44 244
Pudukottai	–	–	911	46 278	911	46 278
Pasumpon	–	–	603	48 734	603	48 734
Kamarajar	5	933	312	17 328	317	18 261
Tirunelveli k.b.	7	1722	1 173	26 683	1 180	28 405
V.O.Chidambarar	1	657	132	1 990	133	2 647
Kanyakumari	4	3184	473	2 993	477	6 177
T.Sambuvarayar	1	2010	–	–	1	2,010
Total	**69**	**58 462**	**8 837**	**300 278**	**8 906**	**358 740**

Table 26: Reservoirs in Tamil Nadu (> 1000 ha)

Sl.No.	*Name of Reservoir*	*Area (ha) (at FRL)*	*Area (ha) (Average)*	*Fish Landing ($t\ yr^{-1}$)*	*Major Fisheries*
Salem District					
1.	Mettur	15346	9324	115	Rohu (19) per cent, Wallago attu (15 per cent), catfishes (14 per cent), *Puntius* spp. (14 per cent), catla (10 per cent)
Periyar District					
2.	Bhavanisagar	7876	3208	179	*L. calbasu* (22 per cent), *L. bata* (19 per cent), *M. aor* (16 per cent), *Puntius dubius* (10 per cent)
South Arcot District					
3.	Veeranam	3885	38	36	Catla (31 per cent), *tilapia* (44 per cent), *rohu* (17 per cent), silver carp (8 per cent)
4.	Perumaleri	2590	1295	–	
5.	Wellington	1554	500	9	Catla(31 per cent), common carp (8 per cent), *rohu* (17 per cent), tilapia (44 per cent)

Sl.No.	Name of Reservoir	Area (ha) (at FRL)	(Average)	Fish Landing (t yr^{-1})	Major Fisheries
Chengalpattu MGR District					
6.	Poondi	3263	1402	15	Tilpia (41 per cent), *common* carp (9 per cent), mrigal (9 per cent), *Puntius* spp. (7 per cent)
Madurai District					
7.	Vaigai	2419	1554	24	Catla (56 per cent), mrigal (22 per cent), silver carp (8 per cent), tilapia (6 per cent)
T. Sambuvarayar District					
8.	Sathanur	2010	56	126	Catla (51 per cent), mrigal (27 per cent), common carp (12 per cent), tilapia (6 per cent)
Kanyakumari District					
9.	Pechiparai	1515	700	9	Puntius spp. (45 per cent), catla(19 per cent), mrigal
Dharmapuri District					
10.	Krishnagiri	1248	768	47	Misc. (73 per cent), tiapia (24 per cent), catla (20 per cent)
11.	Vaniyar	1,093	61	3	Misc. (48 per cent), rohu (27 per cent), catla (13 per cent)

Compiled from: 1. Director of Fisheries, Tamil Nadu; 2. Srivastava, *et al.,* 1985; 3. Anon, 1987.

Table 27: Reservoirs in Tamil Nadu (<1000 ha)

Name of Reservoir	Area (ha) (at FRL)	(Average)	(t yr^{-1})	Fish of Major Fisheries Landing
Chengalpattu MGR District Kolavaieri	900	529	30	–
N. Arcot Ambedkar District Goddar	678	339	3	Tilapia (57 per cent), common carp (25 per cent), catla (10 per cent), rohu (6 per cent)
Dharmapuri District Thumbalahalli	193	89	11	Tilapia (79 per cent), catla (90 per cent), silver carp (4 per cent)
Kasarikulihalla	105	63	4	Tilapia (73 per cent), common carp (13 per cent) mrigal (6 per cent)
Sicclagiri Chinnar	54	27	8	Misc. (58 per cent), tilapia (28 per cent), catla (5 per cent), common carp (5 per cent)
Pambar	243	60	26	Misc. (41 per cent), tilapia (35 per cent), c. carp (8 per cent), rohu (7 per cent)
Barur	256	128	24	Tilapia (90 per cent), misc. (7 per cent), catla (2 per cent)
Nagavathy	118	71	5	Tilapia (76 per cent), *Puntius* spp. (9 per cent), catla (6 per cent)
Thoppaiyar	120	60	15	Tilapia (77 per cent), catla (30 per cent), rohu (5 per cent), mrigal (4 per cent)

Name of Reservoir	*Area (ha)*			*Fish of Major Fisheries Landing*
	(at FRL)	*(Average)*	*(t yr^{-1})*	
Chinnar	170	117	6	Tilaia (26 per cent), *Puntius* spp. (26 per cent), common carp (13 per cent)
Coimbatore District Pillur	400	233	3	–
Tirumoorthy	388	240	13	–
Amaravathy	850	544	112	–
Aliyar	650	384	27	–
Sholaiyar	526	312	5	–
Nirar (upper and lower)	68	35	–	–
Peruvaripallam	290	20	–	–
Nilgiris District Sandynulla	300	156	5	–
Pykara	50	25	3	–
Glenmargam	50	25	1	–
Ooty lake	35	16	4	–
Kundah	40	23	1	–
TR Bazar	11	8	–	–
Dindigul Anna District Kunthirayar	–	–	5	Tilapia (50 per cent), catla (16 per cent), mrigal (9 per cent)
Palarporanthalar	518	376	95	–
Maruthanathi	72	42	8	Tilapia(37 per cent), common carp (24 per cent), catla (23 per cent, mrigal (9 per cent)
Manoor	10	10	2	Common carp (48 per cent), catla (33 per cent), silver carp (19 per cent)
Parappalar	114	60	6	Tilapia (37 per cent), common carp (24 per cent), catla (23 per cent)
Kombai (Peria and Chinna)	35	19	7	Tilapia (71 per cent), catla (14 per cent), common carp (4 per cent), silver carp (4 per cent)
Varathamanathi	80	39	7	Tilapia (71 per cent), catla (14 per cent), common carp (4 per cent), silver carp (4 per cent)
South Arcot District Vidur	798	479	6	Tilapia (55 per cent), common carp (9 per cent), mrigal (9 per cent), *Puntius* sp. (7 per cent)
Gomukhi	360	218	6	*Puntius* sp. (35 per cent), catla (41 per cent), mrigal (11 per cent)
Manimuktha	746	447	12	Catla (47 per cent), common carp (17 per cent), tilapia (6 per cent),
Kamarajar District Vembakottai	467	160	18	Tilapia (22 per cent), mrigal (17 per cent), catla (13 per cent), misc. (35 per cent)
Periyar	76	26	17	Tilapia (53 per cent), mrigal (23 per cent), catla (14 per cent), silver carp (6 per cent)
Kovilar	74	24	12	Tilapia (48 per cent), mrigal (24 per cent), silver carp (11 per cent), catla(10 per cent)

Name of Reservoir	Area (ha)			Fish of Major Fisheries Landing
	(at FRL)	(Average)	(t yr[1])	
Kullur Santhai	316	190	48	Tilapia (70 per cent), mrigal (15 per cent), catla (7 per cent)
Anaikuttam	–	–	1	Tilapia (40 per cent), misc. (34 per cent), mrigal (10 per cent), common carp (9 per cent)
Tirunelveli Kattabomman Dt. Karuppanathi	50	25	4	Tilapia (91 per cent), common carp (4 per cent)
Ramanthi	39	20	4	Tilapia (57 per cent), common carp (22 per cent), *Puntius* spp. (15 per cent)
Gundaru	21	10	1	–
Gadana	80	48	2	*Puntius* spp. (65 per cent), common carp (14 per cent), tilapia (12 per cent)
Manimuthar	940	470	7	*Puntius* spp. (56 per cent), mrigal (16 per cent), tilapia (15 per cent), *Labeo fimbriatus* (6 per cent)
Hope lake	580	300	3	–
Sreemoolaperi	12	10	–	–
Madurai District Manjalar	197	154	28	–
Sathiar	120	60	6	Tilapia (48 per cent), catla (30 per cent), common carp (7 per cent), rohu (6 per cent)
Tiruchirapalli District Kannathudai	50	26	3	Catla (56 per cent), mrigal (22 per cent), tilapia (6 per cent), common carp (12 per cent)
Ponnaniyar	60	16	4	Catla (51 per cent), mrigal (27 per cent), tilapia (6 per cent) common carp (12 per cent)
Uppar	117	60	–	–
Periyar District Uppar	453	405	45	–
Varattupallam	89	53	6	Tilapia (56 per cent), silver carp (17 per cent), catla (17 per cent)
Vattamalai (Karaiodai)	307	66	4	–
Gunderipallem	61	34	15	Silver carp (25 per cent), tilapia (12 per cent), catla (6 per cent)
V. O. Chidambaranar District Kadama	657	328	51	Tilapia (85 per cent), catfish (1 per cent), misc. (13 per cent)
kanyakumari District Chittar-1	293	175	12	*Puntius* spp. (41 per cent), catla (25 per cent)
Chittar-2	414	248		Tilapia (13 per cent), Silver carp (8 per cent)
Perumchani	962	526	9	*Puntius* sp. (71 per cent), tilapia (28 per cent)

5.3. Scientific Investigation of Tamil Nadu Reservoirs

At least twenty reservoirs in Tamil Nadu have been studied and the literature on the subject is rich in descriptive limnology and production dynamics. No other State in the country has generated scientific information of this magnitude on the reservoir ecosystem. The following account provides information on fish stocks and fisheries of 5 major reservoirs of Tamil Nadu.

5.3.1. Stanley Reservoir (Figure 14)

Mettur dam, situated at 11°49'N, is often described as an engineer's delight. A straight gravity structure, 1 615 m long, rising 54 m above the Cauvery river bed, the dam is constructed across two hills of the Eastern Ghats. At the time of its construction, Mettur was the highest masonry dam in Asia and the largest in the world. The dam, constructed primarily to stabilize the irrigation in Thanjavur delta, caters to about one third of the irrigated area of Tamil Nadu, besides generating hydroelectric power of 240 MW. Commenced in 1925, the dam was completed in 1935. Stanley is the largest reservoir in Tamil Nadu.

Stanley reservoir was commissioned in 1939 and it continues to be the only reservoir in the State representing the >10,000 ha category. Situated at an elevation of 243 m above MSL in the Salem district, Stanley reservoir has a waterspread of 15 346 ha and capacity of 2 646 million m^3 at FRL, the average area being 9 324 ha. It receives water from the Cauvery river basin of 42 217 km^2. Morphometric indicators such as high shore development index (6.7) and a volume development index more than unity (1.4) point towards productive nature of the reservoir.

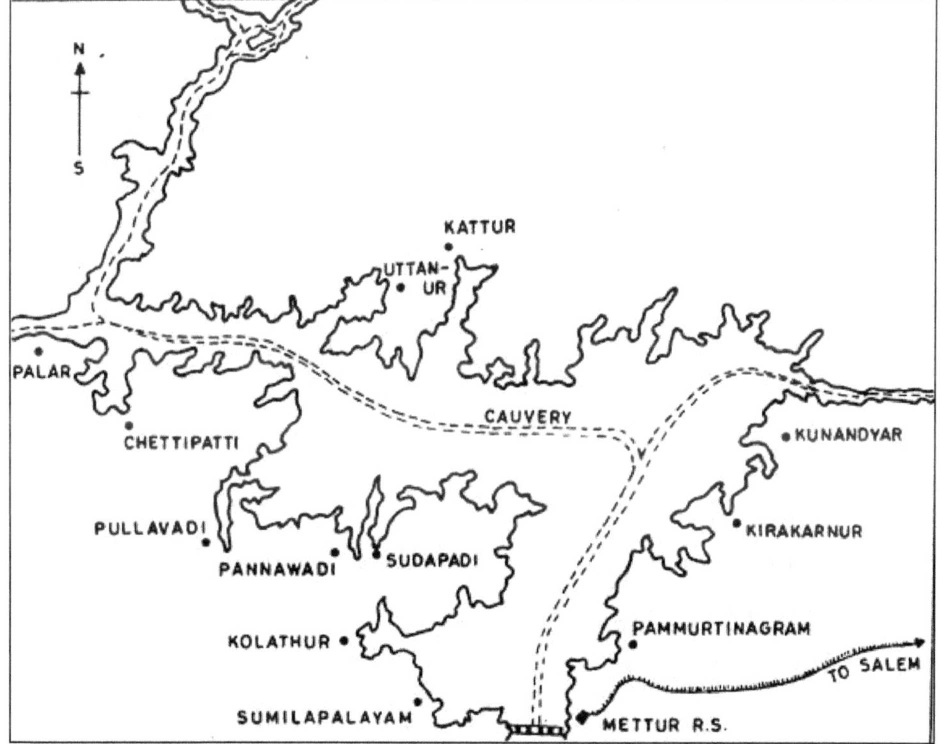

Figure 14: Stanley Reservoir.

Despite being a deep basin, the water column was not stratified thermally, in conformity with the situations reported from elsewhere in peninsular India. Even during the so called winter, air temperature in this part of the country does not drop below 15 °C keeping the water relatively warm. With the onset of summer, when the top layer warms up, there is no cool water below to offer any thermal resistance. Moreover, the release of cooler water from the bottom layer through the outlets of the dam remove any disparity in temperature between the top and bottom layers.

Mettur Dam, Tamil Nadu

Fish and Fisheries

Indigenous ichthyofauna of the river comprised *Acrossocheilus hexagonolepis Puntius carnaticus, P. dubius, Tor putitora, Labeo kontius, L. fimbriatus, Cirrhinus cirrhosa, C reba, Aorichthys aor, A. seenghala, Pangasius pangasius, Silonia silundia* and *Wallago attu* (Chacko *et al.*, 1955). The three downstream barrages on Cauvery *viz.*, the Lower Anicut, the Great Anicut and the Upper Anicut restricted the upriver migration of the anadromous fish, *Tenualosa ilisha* and after the construction of Mettur dam, this fish was reported to disappear completely from upstream stretches of Cauvery. Sreenivasan (1976) reported the disappearance of *Puntius* spp., which used to form 28 per cent of the landings in 1943–44. The indigenous carp *C. cirrhosa* showed an initial increase in catches, but later declined. Low water levels during July in the preceding two or three years were blamed for the spawning failure of *C. cirrhosa. Labeo kontius*, which was common and next only to *C. cirrhosa* oin the Cauvery, also disappeared from the reservoir.

Fishing Gear

Mettur being one among the earliest dams to be commissioned in the State of Tamil Nadu, Stanley reservoir has been considered as a testing ground for evolving reservoir fishing practices in India. The use of gill nets, now used widely in reservoirs all over the country, was developed by the fishermen of Stanley by trial and error methods. They named the floating gill nets as *Rangoon* nets, after the migrant fishermen from Myanmar. This gear is the most suitable for reservoir conditions, inasmuch as most of them have rugged bottom with steep slopes. The fishermen have even evolved the best mesh size to suit each fishery.

5.3.2. Bhavanisagar

Bhavanisagar dam is constructed on the river Bhavani at 11°30'N lat., below its confluence with the river Moyar in the Cauvery basin in the Periyar district. The reservoir has a maximum waterspread of 7 876 ha at FRL which is 280.2 m above the MSL, and its net capacity is 908 million m^3. the average reservoir area is 3 695 ha. During the driest years, the reservoir shrinks to 745 ha in summer. Receiving water from catchment of 4 200 km^2, the reservoir serves irrigation and flood control. The calculated mean depth of Bhavanisagar is 11.5 m, the maximum

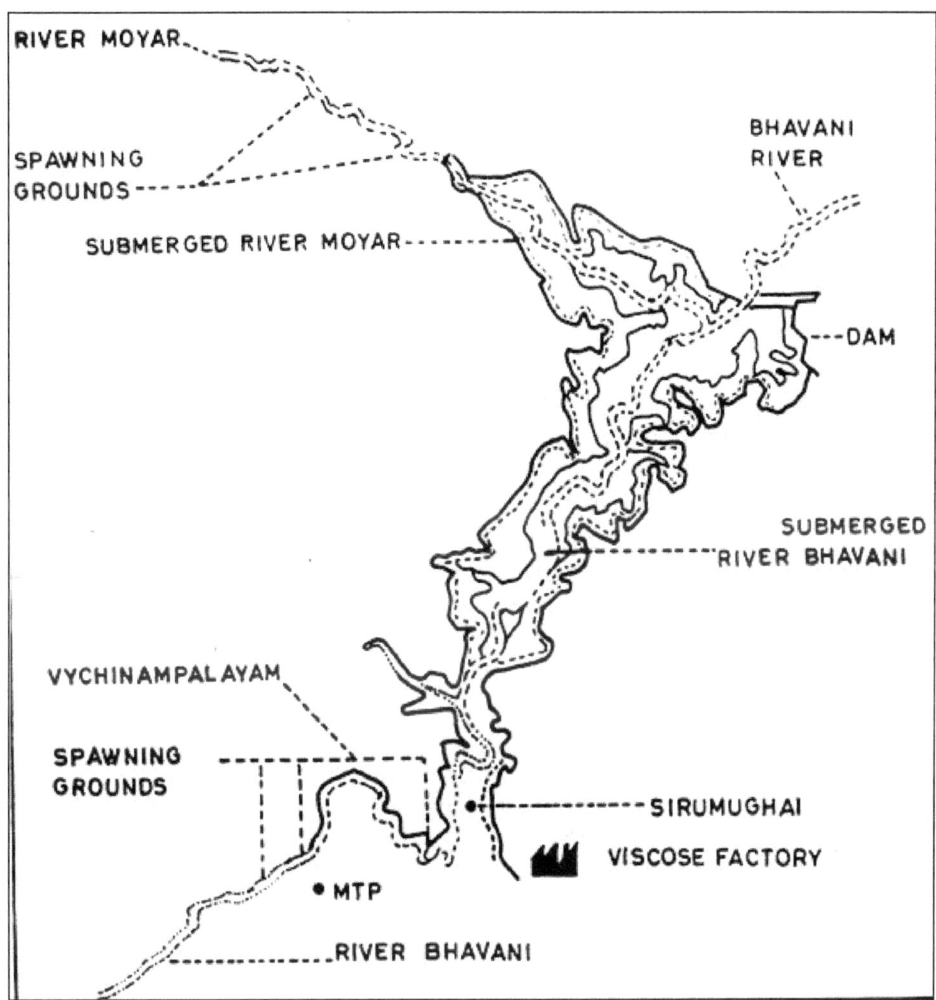

Figure 15: Bhavanisagar Reservoir, Tamil Nadu.

depth being 33 m. The shore development and volume development indices are 4.0 and 1.0 respectively.

Soil of Bhavanisagar reservoir bed is silty clay loam and mostly acidic in reaction. Judged by the presence of available nitrogen, available phosphorus and organic carbon, the soil was categorised by Gupta (1979) as conducive to aquatic productivity. During pre-monsoon months, the pH was found to increase towards neutrality. Water level fluctuations in Bhavanisagar are erratic. In the year 1961 and 1962, annual level fluctuations were of 15.2 and 7.9 m respectively. During 1971 to 1981, water level fluctuations varied from 5.46 m to 22.6 m yr^{-1}, with the accompanied changes in area and capacity of the reservoir. Air temperature regime seems to have undergone a change during the last few years. The maximum air temperature recorded during 1957–62 was 38 °C, while the summer temperature during 1971 to 1981 was an average 40.5 °C. The highest temperature exceeded 45.0 °C during 1974–75 and 1975–76. However, the range of water temperature during 1956–61 and 1971–82 (23.7 °C to 30.2 °C) did not differ much.

The reservoir has a rich plankton community dominated by Cyanophyceae (*Microcystis, Oscillatoria* and *Anabaena*), diatoms (*Nitzschia, Melosira*) and Chlorophyceae (*Pediastrum, Mougeotia*). During the early years (1956– 61), *Nitzschia* and *Melosira* formed an important constituent of the phytoplankton community, along with Cyanophyceae (Sreenivasan, 1964b; Sreenivasan *et al.,* 1964). During 1970's and 80s, however, the diatoms were not very important. Initially, Bhavanisagar had a very high phytoplankton species diversity (Sreenivasan *et al*, 1964; Franklin, 1969). By the 1970s blooms of *Microcystis* (occasionally *Mougeotia*) have become a common feature accompanied by substantial reduction in diversity (Abraham, 1980b). This fall in diversity and increase in dominance of lentic species, aprat from indicating organic enrichment, suggested a transient eutrophication stage through which the reservoir was passing. The rich benthic community of Bhavanisagar was represented by oligochaetes, *Chironomus, Chaoborus* and mayfly nymphs, with a population density of 483 organisms and 3.8 g m^{-2} respectively (Abraham, 1979).

Fisheries

Ichthyofauna of the reservoir comprises 51 species belonging to 11 families (Natarajan *et al.,* 1981). Among them, 18 species contribute to the commercial fisheries. Impoundment did not alter the fish faunistic spectrum to any appreaciable extent as most of the indigenous species (Chacko *et al.,* 1955) except *Tor tor* continue to be caught in the reservoir.

Puntius spp., *T. putitora, T. tor, Accrossocheilus hexagonolepis, Puntius dubius* and *Puntius carnaticus,* along with *Labeo kontius* form the traditional fisheries of river Cauvery and the stretch above the reservoir are believed to be their breeding grounds (Wilson, 1920). One reassuring feature of the reservoir is the continued presence of these indigenous forms especially *P. dubius, P. carnaticus, Labeo kontius, Cirrhinus cirrhosa* and *A. hexagonolepis.* Their survival is attributed to the uninterrupted breeding activities at Moolathurai and Nellithurai, especially in the former when water is released from Pilloor. *P. dubius* which has a fecundity of 66,000, ascends the rivers Cauvery and Moyar during the north east monsoon for spawning and lays eggs in batches of 1,000 to 2,000 on the gravel bed (Ranganathan *et al.,* 1962). It is due to the continued availability of breeding grounds that the indigenous fishes survive in the lake and therefore there is a need to conserve this biologically sensitive environment. *Cirrhinus reba* breeds during June–July at Moolathurai in Bhavani and Hogainakkal in the main river Cauvery (Rao and Gopinathan, 1972). Evidence for recruitment of *Labeo fimbriatus, L. calbasu, L. kontius* and *P. carnaticus* has also been obtained along with successful natural breeding of introduced carps such as catla, rohu and mrigal (Natarajan *et al.*, 1981). Sharp variations in catch per unit of effort in respect of *P. dubius* indicated inconsistencies in annual recruitment (Natarajan *et al.*, 1981). Water level fluctuations seem to be the major factor behind the fish catch fluctuations, rather than fishing pressure.

L. calbasu, which feeds mainly on bottom detritus, is reported to mature at a size of 400 mm. The fish spawns in Moolathurai breeding grounds during south west monsoon (Natarajan, 1971a and b). The mean length in catch varies between 465 and 530 mm among the > 3 year group. Other southwest monsoon breeders are the Indo-Gangetic major carps *Catla catla, Labeo rohita* and *Cirrhinus mrigala*. The indigenous *Puntius dorsalis* breeds intermittently with at least three annual peaks in its spawning intensity (Natarajan *et al.*, 1981). Commercial fish landings from the reservoir during 1964 to 1980 ranged from 85 t to 293 t at an average of 156.5 t. This is rather low, in view of the good primary production rate. Sreenivasan (1976) estimated that 0.0775 per cent of the carbon synthesised and 0.000513 per cent of the solar energy were converted into fish flesh in Bhavanisagar. Community metabolism is mainly through the detritus chain. In the absence of an adequate population of planktiphagous fish, most of the energy at plankton phase is added to detritus. *Labeo calbasu* (95 per cent), *Cirrhinus mrigala* (87.5 per cent), *P. dubius* (94.9 per cent) and *Catla catla* (55 per cent) had a predominantly detritus-based diet. *L. calbasu*, and *C. cirrhosa* are increasing in catches while *P. dubius* and *C. catla* show a decline.

Conservation

Conservation measures to be adopted in Bhavanisagar should be relevant to the feeding, breeding and other habits of the target fishes. *P. dubius* has a natural distribution in the upper reaches of Bhavani and Cauvery in the sub-mountainous Tamil Nadu. This fish feeds mainly on benthic chironomid larvae, oligochaetes, insects and nymphs at the depths of 2 to 10 m. It breeds once a year during October to December, and undertakes a distinct, short upstream spawning migration. The eggs are laid on gravel by the side of rapids and pools.

In Bhavanisagar, the breeding grounds of *P. dubius* are reported to be intact and the fish still thrives in the reservoir. In Stanley, the Hogainakkal fall is a natural barrier obstructing the upstream migration of *P. dubius*. Similarly in Amaravathy, the river stretch above the reservoir has Duvanam Falls where the ready to spawn breeders congregate in large numbers from December. Ranganathan *et al.* (1962) noted that the failure of migration caused atrophy of gonads and breeding failure. Stocking of fry above waterfalls is suggested by the above authors as a conservation measure. However, Sreenivasan (1976) felt that the natural cascades associated with Davanam Falls allowed upstream migration of this fish. Ranganathan *et al.* (1962) recommended stocking of *P. dubius* in reservoirs situated in the sub-montane areas, provided there was a suitable river stretch above and no barrier for upstream migration. Conservation measures should also include:

1. Total closure of fishing activities during the breeding season, and

2. Ban on gill nets < 6 cm in mesh size (stretched) so that fish of three years and less will not be caught.

5.3.3. Amaravathy

The river Amaravathy, a tributary of the Cauvery was dammed in 1958 at 10° 15'N, 5 km below the point, where the river debouches into the plains to create

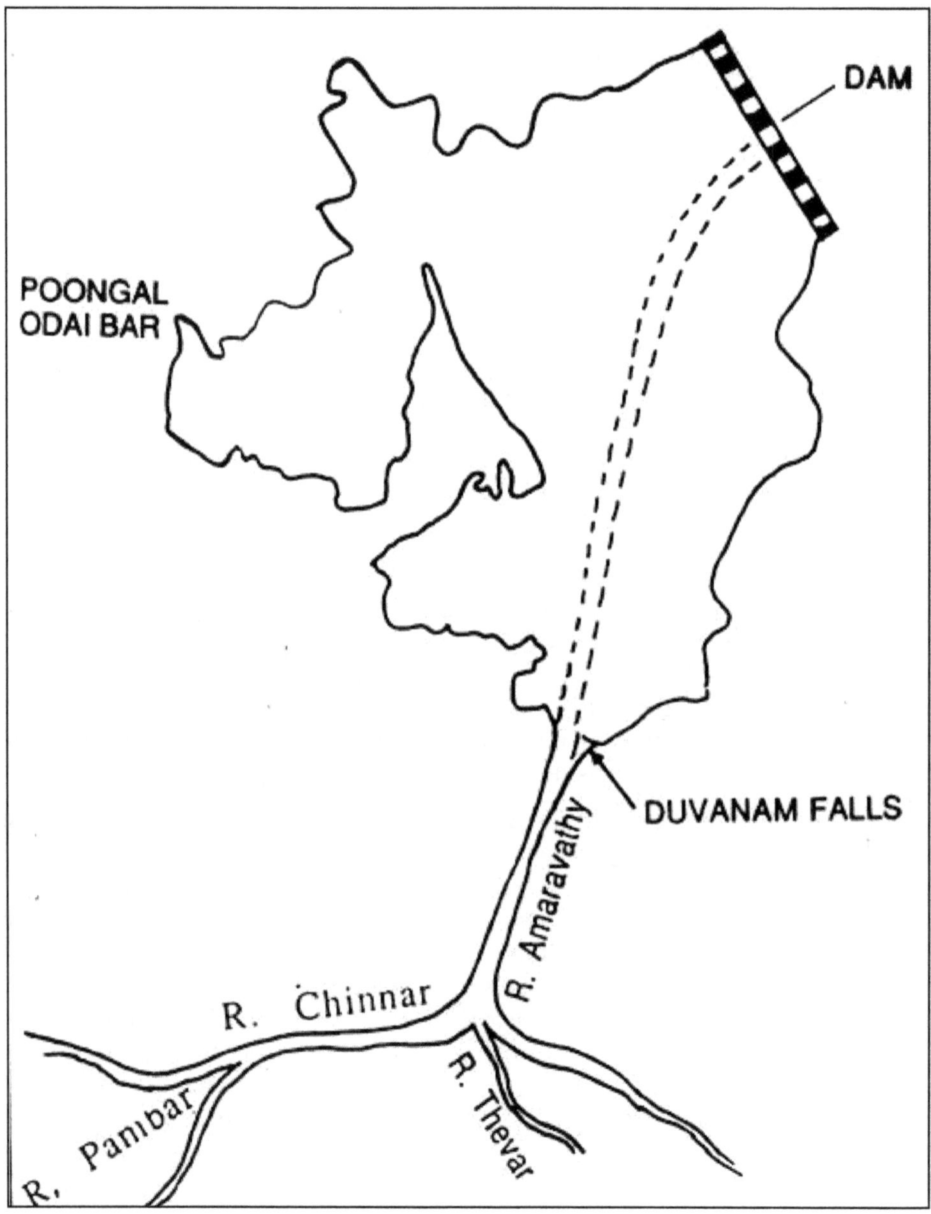

Figure 16: Amaravathy Reservoir, Tamil Nadu.

the Amaravathy reservoir. The reservoir water surface area is 850 ha at the full level of 358 m above MSL. Though small, Amaravathy is one of the most productive reservoirs in the State. The lake has a net capacity of 112.37 million m^3 at FRL and the mean depth is 13.7 m. Volume development and the shore development indices are 1.2 and 2.3 respectively. The catchment area of the reservoir is 832 km^2. The reservoir was described by Sreenivasan (1965a, 1965b, 1976 and 1979a, b, and c.).

Amaravathy has one of the highest primary productivity rates, next only to Tirumoorthy. Carbon fixation rate, measured in terms of oxygen released, was estimated @ 11.320 cal m^{-2} yr^1 × 10^{-6}at a photosynthetic efficiency of 1.442 per cent from sunlight to carbon. An unusually high rate of oxygen release at the rate of 59.1 g m^{-2} day^{-1} has been reported from the reservoir by Sreenivasan (1965a). Plankton community of Amaravathy is dominated by *Microcystis* which blooms round the year.

Alterations in the plankton community due to the reservoir formation is believed to have contributed to radical changes in fish stocks. The indigenous fishes could not thrive on a *Microcystis*-dominated plankton. Almost all indigenous species of commercial value have disappeared. *Puntius dubius*, which ranked first till 1965– 66, forming 55 per cent of the fish landings, dwindled to less than 2 per cent from 1968–69 and disappeared later (Sreenivasan, 1976). *P. carnaticus* was another casualty. Today, the commercial fishery is entirely based on the transplanted fishes including some exotic species. Tilapia, *Oreochromis mossambicus*, which was stocked in the year 1957–58, gradually established a foothold any by mid-seventies, they formed more than 90 per cent of the catch. The common carp, *Cyprinus carpio* has increased since 1975–76.

Performance of tilapia in Amaravathy, over the years was phenomenal. This allayed the fears of stunted growth, a common complaint against this particular tilapia (*Oreochromis mossambicus*) in pond environment. Moreover, the fishery based on tilapia could effciently convert energy into fish flesh. The energy conversion rates from primary producer to fish and light energy to fish were 0.2081 per cent and 0.003003 per cent respectively, but being very high and next only to Sathanur reservoir.

5.3.4. Sathanur

Sathanur reservoir was created in 1957 on the River Ponniar at Sathanur Village in Tiruvannamalai Sambuvarayar district (12° 12'N). It covers 2,010 ha (Figure 17) at the FRL of 222.2 m. The reservoir has a capacity of 228.91 million m^3 at full level and a mean depth of 11.4 m. Shore and volume development indices are favourable for productivity. Sathanur is a hard water reservoir (total hardness range from 112–254 mg l^{-1}), with a matching high specific conductivity (320-800 umhos), and total alkalinity (145–616 mg l^{-1}). Values of all the three parameters mentioned above are some of the highest recorded in Tamil Nadu reservoirs.

Both morphometric as well as metabolic indicators point to the productive nature of the reservoir There is a high rate of carbon fixation. The primary chemical

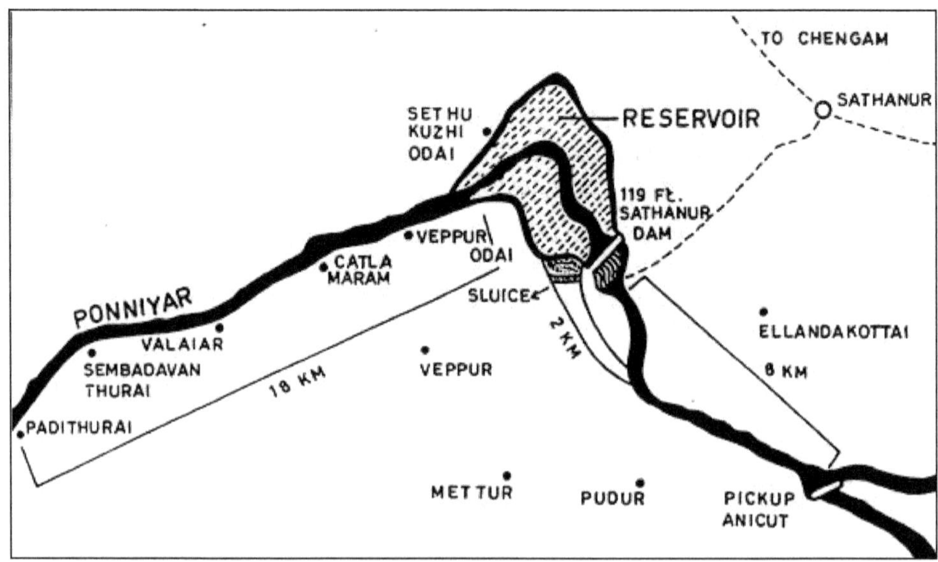

Figure 17: Sathanur Reservoir, Tamil Nadu.

energy fixed by producers is estimated at 8.1 × 10^6 cal $m^{-2}yr^{-1}$ at a photosynthetic efficiency of 1.047 per cent. The fish, *Wallago attu, Cirrhinus reba, Labeo kontius, L. fimbriatus, Notopterus notopterus* and *Mastacembelus armatus* are indigenous species, while *Catla catla, Cirrhinus mrigala, C. cirrhosa, L.calbasu, L. rohita* and *Cyprinus carpio* (Bangkok strain) are introduced.

Drifting gill nets of mesh bar 5, 6 and 7.5 cm are employed for fishing. A fishing unit comprises 10 gill nets, each 50 m in length, a coracle (a saucer-shaped country craft made of split bamboo and hide) and two men. During this period, catla has established firmly in the reservoir. The indigenous *L. fimbriatus*, which was dominant (36 per cent), represented only 3 per cent of the total catch. *W. attu*, the main predator, is less in percentage, though steady in its occurrence. Presently, catla along with other Indian major carps form more than 78 per cent of the catch.

It is encouraging to note that *C. catla*, the prime fish species of Sathanur, is represented in the catch by the 3 to 4 year class. The presence of W. attu has not affected the carp fishery. Prabhavati and Sreenivasan (1979) believe that small numbers of this predator are welcome as a means to check the weed fish populations. The decline in populations of *L. fimbriatus* and *L. calbasu* is attributed to their breeding failure. Until 1977, fish catches in Sathanur were steadily increasing every year, despite a steady increase in fishing effort (Sreenivasan, 1979a). The catch per unit effort (as described by Rangathan and Venkataswamy, 1967) still remained high at 7 318 to 14 295 kg yr^{-1}, thus suggesting scope for further increase in effort.

5.3.5. Aliyar Reservoir (Figure 18)

Aliyar dam was constructed in 1962 across the Aliyar river. Aliyar reservoir, situated between 10°15' and 10°30' N and 76°50' and 77° 10' E, covers 646.0 ha at the FRL of 320.04 m above MSL. Apart from a catchment area of 468.8 km^2, the lake receives water from Upper Aliyar reservoir through a hydro-electric power station at Navamali and the Parambikulam reservoir (Kerala) through a contour canal. The reservoir with its maximum depth of 36.5 m and mean depth of 18.2 m, exhibits annual water level fluctuations ranging from 13.15 m to 31.03 m at an average of 21 m per year.

Morphometric Features of Aliyar Reservoir

Catchment area	:	468.8 km^2
Length of shoreline at FRL	:	16.0 km
River bed level (MSL)	:	283.6 m
Full reservoir level (MSL)	:	320.04 m
Maximum depth	:	36.58 m
Dead storage level	:	293.0 m
Mean depth	:	18.2 m

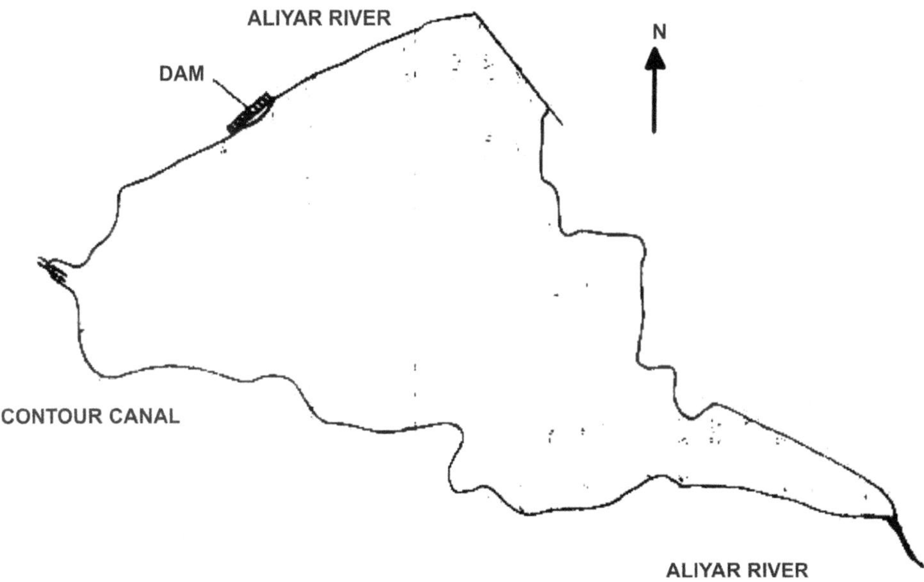

Figure 18: Aliyar Reservoir, Tamil Nadu.

Average water level		
Fluctuation	:	21.0 m
Area of dead storage level	:	2.5 ha
Maximum area at FRL	:	646.0 ha
Average area (DSL + FRL)/2	:	324.25 ha
Capacity at FRL	:	10 942.8 ha m
Volume development	:	1.2

Source: CICFRI, Barrackpore.

Common Planktonic Organisms in Aliyar Reservoir

Chlorophyceae	:	*Mougeotia sp., Pediastrum sp. Staurastrum sp.*
Myxophyceae	:	*Microcystis sp., Anabaena sp.*
Bacillariophyceae	:	*Asterionella sp., Synedra sp., Navicula sp.*
Dinophyceae	:	*Ceratium sp.*
Protozoa	:	*Actinospherium sp., Cyclotella sp.*
Rotifera	:	*Keratella sp., Brachionus sp., Filinia sp.*
Cladocera	:	*Daphnia sp.*
Copepoda	:	*Cyclops sp., Diaptomus sp., nauplii*

Numerically, the plankton population is dominated throughout the year by *Microcystis* sp. The only other biota worth mentioning are *Mougeotia*, the filamentous alga, *Cyclops* (2.9 per cent) and *Keratella* (2.5 per cent). The diversity of species in the reservoir is poor with negligible sectoral variation. The year-round distinct dominance of *Microcystis* suppresses development of a species-rich plankton population in the reservoir.

Benthos

Bottom macrofauna : The average standing crop of bottom macrofauna in Aliyar, recorded for four years from the depths of 2 to 10 m, is estimated at 258 ind. m^{-2} (3.18 g m^{-2}). Oligochaetes dominate both in terms of numbers (47.8 per cent) and weight (86.8 per cent). Shallows have a low density of macrobenthos and their concentration generally increases with depth. The maximum concentration of benthic organisms (2 331 organisms m^{-2}/48.45 gm^{-2}) was recorded at 30 m depth.

Fish Fauna

The indigenous ichthyofauna of the reservoir consists of 40 species belonging to 13 families, besides the seven stocked species.

Ichthyofauna of Aliyar Reservoir

Family : CYPRINIDAE

Sub-family : Cyprininae

1. *Catla catla* (Hamilton-Buchanan).
2. *Cirrhinus mrigala* (Hamilton-Buchanan)
3. *Ctenopharyngodon idella* (Valenciennes)
4. *Cyprinus carpio communis Linnaeus*
5. *Labeo rohita* (Hamilton-Buchanan)

Sub-family : Leuciscinae

6. *Hypophthalmichthys molitrix* (Valenciennes)

Family : CICHLIDAE

7. *Oreochromis mossambicus* (Peters)

Family : NOTOPTERIDAE

8. *Notopterus notopterus* (Pallas)

Family : ANGUILLIDAE

9. *Anguilla bengalensis bengalnesis* (Gray)

Sub-family : Cyprininae

10. *Cirrhinus cirrhosa* (Bloch)
11. *Cyprinus carpio* var. specularis Lacepede
12. Puntius *dubius* (Day)
13. *Labeo boga* (Hamilton)
14. *Labeo calbasu* (Hamilton)
15. *Labeo fimbriatus* (Bloch)
16. *Labeo kontius* (Jerdon)
17. *Puntius carnaticus* (Jerson)
18. *Puntius chola* (Hamilton)
19. *Puntius densonii* (Day)
20. *Puntius dorsalis* (Jerdon)
21. *Puntius filamentosus* (Valenciennes)
22. *Puntius mahecola* (Valenciennes)
23. *Puntius sarana* (Hamilton)
24. *Puntius ticto* (Hamilton)
25. *Puntius punctatus* (Day)

26. *Puntius melanampyx* (Day)
27. *Tor khudree malabaricus* (Jerdon)
28. *Salmostoma chela untrachi* (Day)

Sub-family : Rasborinae

29. *Amblypharyngodon melettinus* (Val.)
30. *Barilius gatensis* (Valenciennes)
31. *Danio aequipinnatus* (McClelland)
32. *Perluciosoma danicontus* (Hamilton)

Sub-family : Garrinae

33. *Garra mecclellandi* (Jerdon)

Family : BAGRIDAE

34. *Mystus malabaricus* (Jerson)

Family : SILURIDAE

35. *Ompok bimaculatus* (Bloch)
36. *Ompok malabaricus* (Valeniciennes)

Family : SISORIDAE

37. *Glyptothorax housei* Herre

Family : CLARIDAE

38. *Clarias batrachus* (Linnaeus)

Family : HETROPNEUSTIDAE

39. *Heteropneustes fossilis* (Bloch)

Family : BELONIDAE

40. *Xenentodon cancila* (Hamilton)

Family : NANDIDAE

Sub-family : Pristolepidinae

41. *Pristolepsis* (*malabaricus*) *marginata* (Jerdon)

Family : CICHLIDAE

42. *Etroplus canarensis* (Day)
43. *Etroplus maculatus* (Bloch)

Family : GOBIIDAE

44. *Glossogobius giuris* (Hamilton)

Family : CHANNIDAE

45. *Channa marulius* (Hamilton)

Family : MASTACEMBELIDAE

46. *Macrognathus guentheri* (Day)
47. *Mastacembelus armatus* (Lacepede)

No. 1 to 7 - Species stocked

5.3.6. Fisheries Management

Current fish yield rate of Aliyar reservoir is one of the highest in the country. This reservior acted as testing ground for the field trials of the scientific management package developed by the CICFRI and the yield optimisation achieved thereof is a standing testimony to the validity of the package. The State Fisheries Department launched a stocking programme in the reservoir immediately after its formation in 1962 (Selvaraj and Murugesan, 1990). The stocking density was erratic in the earlier years with a high preponderance of medium and minor carps including*Puntius carnaticus, p. dubius, Labeo Kontius, L. fimbriatus, L.calbasu, Cirrhinus reba, C. cirrhosa, Acrossocheilus hexagonolepis* and *Tor khudree malabaricus. Apparently,* availability of fish seed in the vicinity and other logistics of seed supply governed the stocking density and the species-mix, rather than any ecological considerations. The size of fry stocked was usually so small that they fell an easy prey to predators.

The exploitation system followed during the early phase also left much to be desired. The fishing was done, both by the Department of Fisheries and private fishing units, on a royalty basis, without any restriction on mesh size. This resulted in indiscriminate exploitation of stocked fishes much below the desirable size. A cumulative effect of a number of irrational reservoir management practices resulted in very low yields, ranging from 2.67 to 54.7 kg ha ^{1}, at an average of 26.21 kg ha^{1} yr^{1}.The reservoir was subjected to scientific investigations since 1983 and a number of measures were initiated to develop the fisheries. The results of scientific management started reflecting in the yield rate from 1985. The management options used for enhancing fish yield were:

1. Stocking was limited to Indian major carps
2. Size at stocking was increased to advanced fingerlings (>100mm)
3. Stocking density was reduced to 236–312 ha^{1} (Av. 637 ha^{1})
4. The stocking schedule was spread over different months of the year
5. A strict mesh regulation, banning nets < 50 mm in mesh bar (knot to knot),
6. Ban on catching Indian major carps less than 1 kg in size.

5.3.7. Status of Reservoir Fisheries in Tamil Nadu

Tamil Nadu is harnessing its water resources to the fullest extent by creating impoundments all over the State and this has resulted in the location of reservoirs in a variety of places ranging from the uplands of the west to the coastal deltas in the east. They are situated in diverse morpho-edaphic and geo-climatic environs, receiving drainage from varying types of catchment areas. All these factors add an element of diversity to the environmental conditions under which the reservoirs exist. Scientific data on various environmental variables and biotic communities are available on at least 21 reservoirs in the State. Sreenivasan (1976) concluded that the production processes and the community succession in these reservoirs are so complex and interrelated that classifying them, based on the conventional methods of morphometric and edaphic production indices, was almost impossible. Most of the limnological information with predictive value has been found to be of limited application in these reservoirs.

Most of the man-made lakes in Tamil Nadu, except Pykara, Sandynulla and Mukerti, are situated in medium to low altitude, the elevation ranging from 38 to 483 m above MSL. Thus, almost all the reservoirs are exposed to the warm tropical climate with air temperature never dropping below 15°C. A high temperature regime throught the year, coupled with plentiful sunshine has compensated for much of the adverse attributes in the shallow reservoirs of Tamil Nadu, which explains the non-conformity with morphometric and chemical indicators of productivity. From the two mountain reservoirs *viz.*, Sandynulla and Pykara, situated at +2,120 and +2,036 m altitude respectively, the former is organically rich and highly productive, harbouring a permanent bloom of *Microcystis*, whereas the latter is truly oligotrophic, with low alkalinity, weak oxygen stratification and an aidic water with the presence of dissolved carbon dioxide even at the top layer (Sreenivasan, 1968). Manimuthar and Pechiparai, two small reservoirs in the plains remained oligotrophic mainly due to nutrient-starved catchments.

Shallow lakes are often considered to be more productive, as larger proportion of the water volume remains in the euphotic zone. This maxim does not hold good in respect of Tamil Nadu reservoirs. Neither mean depth nor the maximum depth has any correlation with productivity. Vidur reservoir, despite being the shallowest of all reservoirs (mean depth of 2.1 m), did not develop a rich plankton community, even after seven years of its impoundment. Deeper reservoirs like Amaravathy (13.7 m), Aliyar (16.8 m) and Tirumoorthy (11m), on the other hand, developed algal blooms as soon as the dams were sealed. Other morphometric indices like shoreline development and volume development also did not convey the status of productivity with any degree of accuracy. Sreenivasan (1976) examined the area, mean depth and volume, of a number of reservoirs in Tamil Nadu and found no correlation among the above variables and productivity.

Total dissolved solids, measured in terms of specific conductivity or the bicarbonate alkalinity, is an index of biogenic production potential. Eventhough this indirect measure of edaphic quality reflected the status, to an appreciable extent, there are many exceptions. For instance, Amaravathy and Tirumoorthy, despite their low values of total alkalinity, specific conductivity and hardness, have high organic productivity.

Thus, the productive nature of each reservoir is determined by different factors. For example, the low productivity of Manimuthar and Pechiparai is due to morphometric features (deep basin and low catchment area respectively), while the high rate of productivity in Sandynulla is due to the rapid eutrophication from urban wastes. However, a close examination of the reservoirs reveals certain definite indicators of productivity. The nutrients load of the inflowing waters being dependent on the nature of catchment, quality of the catchment area plays an important role in determining the nutrient status of a reservoir. The percentage of cultivated area is often taken as a criterion for assessing the catchment quality. The high bicarbonate alkalinity and hardness of Sathanur, Krishnagiri and Vidur reservoirs can be directly attributed to the geochemical status of its catchment (Sreenivasan, 1976). Similarly, the low values of alkalinity, conductivity and hardness of Hope lake, Manimuthar, Pechiparai and Perumchani are a reflection of the poor nutrient status of their catchment area.

Thermal regime and oxygen distribution, determining the vertical profile of chemical parameters give a more reliable measure of productivity status in the reservoirs under study. The chemical stratification is the measure of metabolic and organic processes and gives a more direct indication about the trophic status. Absence of proper thermocline is the characteristic feature of all reservoirs in the state. In many cases, the inflowing river water is much cooler than the standing water, as the former flows down form heavily wooded sheltered forests. This cooler incoming water, unlike in the temperate region, readily mixes with the lentic water. Other reasons for the ready mixing are the withdrawal of water from the bottom and the wind-induced turbulence.

Oxygen depletion at the bottom is a very reliable indicator of productivity in tropical waters, especially because it is unaccompanied by thermal stratification. The bacterial decomposition of organic matter at the bottom is hastened by high temperature. Similarly, the photosynthetic processes at the euphotic zone keep up a steady release of oxygen. All the productive reservoirs in the State have the klinograde oxygen distribution, while the oligotrophic Pykara, Uppar and Pambar reservoirs have the orthograde oxygen distribution. Where oxygen depleted water is discharged downstream, it gets reoxygenated within 100 meters downstream (Sreenivasan, 1970a) and its impact on riverine aquatic communities is negligible.

It is often believed that total alkalinity above 50 (Northcote and Larkin,1956) and 90 mg l^{-1} (Spencer, 1964) is indicative of high productivity. In Amaravathy, the inflowing water is poor in total alkalinity and hardness. However, the standing crop of plankton provides a continuous plankton rain to enrich the bottom, which in turn decomposes to release the nutrients and keep the carbon production going. Thus, it is the vertical gradient of these chemical parameters, rather than the ambient concentration in inflowing water, that indicates the status of productivity.

Changes in Biotic Communities

Construction of dams and the subsequent impoundment brings in drastic changes in the biotic communities at all levels. These changes at the micro-level have far reaching impact on the trophic structure and events of the ecosystem, affecting the survival of higher organisms, especially fishes. Sandynulla reservoir, despite its deep basin and high altitude, sustains a rich plankton community, represented mainly by *Microcystis* in blooms. Drastic reduction in diversity of planktonic organisms with a high index of *concentration of dominance* is characteristic of reservoirs which receive rich supply of nutrients. Most of the reservoirs in the state such as Stanley, Bhavanisagar, Amaravathy, Aliyar, Tirumoorthy, Vaigai, Vidur and Poondi have blooms of blue-greens round the year.

Apart from light penetration and nutrient loading, water renewal plays an important role in the abundance of planktonic organisms. Abraham (1980b) pointed out that plankton density in Bhavanisagar reservoir was rich during the months of low level fluctuations. In reservoirs with violent level fluctuations, the annual flood water sweeps away not only the nutrients but even physically dislodges all the passive communities like plankton from the system. It is after the closure of outlets that the plankton community starts developing again. The same is true with benthic and macrophytic communities.

Fish Fauna

Fish fauna of the reservoirs in Tamil Nadu is comprised indigenous species, those transplanted from other river systems and the exotics. In most cases, the last two categories support commercial fisheries while the indigenous ones are on the decline. The gradual phasing out of indigenous fishes is a matter of concern. Reasons for their decline can be summarised as:

1. Loss of habitat,
2. Changes in fish food organisms
3. Obstruction of migratory pathways,
4. Loss of breeding grounds, and
5. Lossing out in competition with the transplnted species.

The anadromous Indian shad (*Tenualosa ilisha*), the fishery of which has collapsed in the Cauvery above the Mettur Dam, is the most serious loss of indigenous fishes. The hilsa from Cauvery-Coleroon system used to ascend the River Cauvery for breeding. The three lower barrages *viz.*, the Upper Anicut, the Grand Anicut and the Lower Anicut restricted hilsa runs even before the Mettur dam came up. After commissioning the dam, hilsa disappeared from the upstream stretches.

Another species that should attract the attention of conservationists is *Puntius dubius*, indigenous to Mettur, Bhavanisagar, Amaravathy and Krishnagiri reservoirs. In Amaravathy, it contributed 55 per cent of the catch in the mid-960s and declined to 0.15 per cent in the 1970s and thereafter disappeared completely. In Bhavanisagar, it still constitutes 5 to 10 per cent of the total catch. Decline of *P. dubius* in the reservoirs is mainly due to recruitment failure. The genus *Puntius* had the maximum species diversity in the region which, in addition to 5 commercially important species, comprised a number of uneconomic fishes.

Mahseers, comprising *Tor tor, T. putitora* and *Acrossocheilus hexagonolepis* are also threatened. *Cirrhinus cirrhosa*, althought still caught in Bhavanisagar, Stanley and other reservoirs is on the decline. This fish faces severe competition from C. *mrigala* that has been introduced into the reservoirs. Tilapia *Oreochromis mossambicus* has been the most successful among the introduced fishes. At least 24 reservoirs in Tamil Nadu have tilapia either as the major component of fish catch or represent a dominant fishery as in Amaravathy, where tilapia outnumbers other fishes by 90 to 10. Catla is the major fishery only in five reservoirs*viz.*, Sathanur, Ponnaniyar, Kannathudai, Manimukta and Gomukhi. The indigenous fishes still dominate in Stanley, Bhavanisagar (*P. dubius*), Manimuthar, Gadana, Chittar and Perunchani (*Puntius* spp.).

Fish Production

Sixty-nine reservoirs of Tamil Nadu produce 1 323 t of fish annually. Figures from the small reservoirs designated as *irrigation tanks* are not available. Small (<1,000 ha in size) reservoirs' share to this fish production is 760 t. Thus 57.4 per cent of the production comes from small reservoirs which form just 26.8 per cent of the total area. Production from unit area is also high in respect of the small reservoirs. The small reservoirs yield 48.5 kg ha^{-1}, calculated for FRL, 88.01 kg ha^{-1}, if calculated for mean area. Medium reservoirs have fish yield of 13.74 kg ha^{-1} (FRL) and 42.20 kg ha^{-1}(mean surface area). Large reservoirs have the lowest yields of 12.66 and 23.46 kg ha^{-1}, calculated at FRL and mean surface area respectively (Table 20). If all categories of reservoirs are pooled together, the Tamil Nadu reservoirs produce 22.6 kg ha^{-1} (at FRL) and 48.04 kg ha^{-1} (at mean surface are level).

Table 27: Production and Yields from different Categories of Reservoir in Tamil Nadu

Category	*Size (ha)*	*Total Fish Catch (t)*	*Yield (kg ha^{-1})*	
			At FRL Area	*At Mean Surface Area*
I	<1,000	760	48.50	88.10
II	1,000–5,000	269	13.74	42.20
III	>5,000	294	12.66	23.46
Total/Mean	–	1323	22.63	48.04

Data pertain to 1993.

Source: Govt. of Tamil Nadu.

This is one of the highest yields in the country, but yields vary widely among the districts. Dindigul Anna district produces 156.81 kg ha^{-1} from its 130 reservoirs. Thiruvannamalai Sambuvarayar, Kamarajar and Coimbatore districts have average yields rates above 50 kg ha^{-1}.

Fish Yield Potential

Fish yield potential, based on the primary production rates, has been estimated in respect of seven reservoirs (Sreenivasan, 1976). Tirumoorthy is the most efficient in converting solar energy (1.803 per cent), followed by Amaravathy (1.442 per cent) and Bhavanisagar (1.050 per cent). The energy harvested as fish is estimated at 0.02931 per cent to 0.2152 per cent of the energy fixed by the primary producers. The best conversion of Sathanur is 0.2152 per cent and considering the fact that up to 1 per cent of the carbon can be harvested, there is sufficient room for increasing the fish yield. In Amaravathy, there is scope for a five-fold increase in the fish production. The mono-species fisheries of catla and tilapia in Sathanur and Amaravathy facilitate better conversion of primary production to fish flesh and consequently, these two species are better converters of energy.

Increase in Productivity

Nearly 60 per cent of the reservoirs in the State of Tamil Nadu and all the tanks are less than 500 ha in size, making them ideal for extensive aquaculture. Management strategy in very small reservoirs and the irrigation tanks should centre around an imaginative stocking and harvesting schedule to allow the stocked fish maximum time to grow. In many cases, the reservoirs nearly or completely dry up during summer, eliminating chance for fish reproduction. This calls for *put* and *take* system of management, where the selected species of fishes are stocked and harvested, all within less than one year. Such systems can also be fertilized to enhance plankton production. The small reservoirs also offer opportunities to integrate aquaculture with animal husbandry and livestock farming. However, the possible conflicts with other water users may have to be reconciled.

Pollution

The reservoirs in Tamil Nadu do not face problems from pollution on a very large scale. In the absence of any organised sewage disposal system in small towns. The community wastes often find their way to the reservoir. From fisheries point of view, this organic loading within certain limits is acceptable. Nevertheless, the problem needs to be addressed from a public health and aesthetic point of view. Organic wastes pose serious eutrophication problems in Sandynulla, which receives city sewage from Ooty. Non-point sources, mainly agricultural wastes, are potentially harmful. Accumulation of pesticide residues in water or sediment phase, their bioaccumulation in organisms and biomagnification in different trophic levels have been not yet investigated in Tamil Nadu and need urgent attention.

Some such instances of pollution from industrial sources have been reported from many reservoirs with varying degrees of impact on the environment and biotic communities. Fish mortalities in Mettur dam area due to effluent discharge from the Mettur Chemical and Industrial Corporation were reported as early as 1949. The effluents contained high concentrations of dissolved and insoluble solids, chlorides and free chlorine and were highly alkaline. However, the effluents were dangerous only during low discharge from the dam when the offensive fluids remained entrapped in rock pools where they affected fish. Apart from this, a series of chemical plants at Mettur discharge their effluents into the Ellis surplus side, but none of them is known to affect the fish life downstream. Waste discharge into the Bhavani river from the Seshasayee Pulp and Board Limited, Pallipalayam, manufacturing writing paper and boards using bamboos. Wood and bagasse adopting the sulfate processes, has been studied and found to be innocuous by Sreenivasan (1979a and b). The South India Viscose, manufacturing viscose rayons and staple fibre, discharges 16,000 m^{3}of wastes into the river Bhavani at Sirumughal every year. The sulfite process of manufacturing is highly polluting. There have been reports of fish mortality in Bhavanisagar reservoir due to this effluent discharge. Low dissolved oxygen level and high carbon dioxide concentrations were recorded in affected areas. Combining effluents with the septic wastes from the residential colony and treatment of the combined wastes in stabilization ponds before discharging them into river have been suggested by Sreenivasan (1979b).

Exotic Fishes

Among the fishes transplanted into reservoirs of the State, the tilapia, *Oreochromis mossambicus* has been the most successful. It has become so ubiquitious that no water body in Tamil Nadu is free from it. After the initial enthusiasm with which the people have accepted it, the fish has become very unpopular due to the stunted size. In ponds, myriad of tilapia flocking the entire water body is a very common sight. The small-sized tilapia does not fetch a remunerative price and only way to utilize them is conversion into fish meal. However, in the reservoirs, the problem of stunted growth has not yet arisen. In Mettur, where the fish forms the bulk of the catch, the individual size is more than 0.68 kg, which is readily accepted by the market. However, it is significant to note that the average size of tilapia decreased from 1.5 kg in 1968 to 0.68 kg over a period of 25 years. Excessive proliferation of tilapia in reservoirs could be kept under check by predators like *Notopterus cinotala. Wallago attu, Channa spp.* and eels, and frequent level fluctuation that exposes the breeding pits.

After heavy stocking of tilapia in Vaigai, Sathanur, Amaravathy, Krishnagiri and Manimuthar reservoirs in the early 1960s, fairly large sized fish were caught in gill nets. In Vaigai they formed 25–50 per cent of the catch by weight soon after stocking. Presently *O. mossambicus*is the single largest component of the fishery in the reservoir. Opinion is divided on the utility of the fish for yield optimization in Indian waters.

Impact of Stocking on the Species Spectrum and Productivity

Catla catla, Cirrhinus mrigala, C. cirrhosa, Labeo rohita, L. calbasu, L. fimbriatus and *Cyprinus carpio* are the species being stocked in reservoirs. In Mettur, catla started appearing in the fisheries in 1943–44 and gained the third rank in ten years to form 16 per cent of the catch. During the next decade, *i.e.*, from mid 1950s to 1960s catla was the most dominant component of the catch. It had a fluctuating fortune, since then. Today, catla form nearly 10 per cent of the catch in the reservoir. This is the fate of catla in most of the reservoirs, where it does not reproduce. The exception is Sathanur, where self-reproducing catla contributes up to 91 per cent of the total catch, while the fish forms only 10–20 per cent of the catch in many reservoirs with regular stocking. *C. mrigala* and *L. rohita* are the other two species which could not establish themselves in any appreciable extent. *L.calbasu* was successfully introduced in Stanley and Bhavanisagar.

In most of the reservoirs, stocking is practised on an ad-hoc basis. Introduction of new species should be done with due consideration of long-term and short-term objectives. In larger reservoirs, the first option should be to stock suitable fishes which have a reasonable chance to establish self reproducing stocks. The main criterion for selection should be to fill a vacant or underutilized niche. On the other hand, a small reservoir with no breeding facilities needs to be managed on a continuous stocking and harvesting basis. This is a culture-based fishery, where what is stocked is harvested. The stocking material is selected after an assessment of trophic structure *i.e.*, fish feed availability. Such approach is adopted successfully in Aliyar.

The organically rich Aliyar reservoir has a permanent bloom of *Microcystis* which is not directly consumed by any species of fish, but it is added to the detritus. The detritivorous *Cirrhinus mrigala* (37 per cent) and *Cyprinus carpio* (24 per cent) together form 61 per cent of the catch. followed by catla (21 per cent) and *Labeo rohita* (16 per cent). The Aliyar experience stressess the importance of utilising the detritus as a fish food resource, where direct plankton feeders are not available or cannot utilize blue-green algae. Job and Kannan (1980) corroborated this by studying the caloric value and chlorophyll content in detritus phase in Sathiar reservoir, where the introduced carps account for 98 per cent of the total fish harvested from the lake. In both reservoirs, the fisheries is totally dependent on the stocked fish. The stocking and harvesting schedule is carefully planned and the fish is removed after attaining a certain size. Smaller fish caught accidentally are released back into the reservoir. This management approach has increased the yield in Aliyar from the initial 35 kg ha^{-1} up to 192 kg ha^{-1}.

Craft and Gear

The most common gear used by the fishermen in the reservoirs of Tamil Nadu is the gill net of entangling type. The *Rangoon* net is the most popular among them. which has been introduced into the state from Andhra Pradesh (Nayar, 1979). Another variety of gill net used extensively in reservoirs is *udu vala*, a narrow bottom set gill net, operated usually in the shallow areas of the reservoir. Gill nets of varying mesh sizes are used to catch different species of fish. The small–meshed nets are used for *Oxygaster phulo, Osteobrama vigorsii* and other small fish. Large–meshed nets are called *catla nets* in Tamil Nadu.

There is no uniform standard length or height for gill nets. For the sake of computing catch per unit of effort in Bhavanisagar, Ranganathan and Venkataswamy (1966) reckoned 20 pieces of *Rangoon* nets comprising three different mesh sizes and a coracle as a standard unit. Mesh sizes were 6.3 cm (50 m length), 5 cm (40 m length) and 10 cm (100 m length). The All India Coordinated Project on Reservoir

Fisheries has considered 50 m hung length as a unit and the effort was identified in terms of cluster of 50 m units.

Apart from gill nets, simple long lines, pole and line and cast nets (both stringed and unstringed) are occasionally employed. Common coracle, a saucer shaped country craft made of split bamboo mat covered with buffalow hide is the most popular craft in reservoirs.

Table 28: Salient Features of some Reservoirs in Tamil Nadu

Parameters	*Stanley*	*Bhavanisagar*	*Poondi*	*Vaigai*	*Sathanur*	*Krishnagiri*
Year of sealing	1939	1953	1957			
Area at FRL (ha)	15 346	7 876	3 263	2 419	2,010	1 248
Area at DSL (ha)	–	120	–	–	–	–
Average area (ha)	–	3998	–	–	–	–
Volume (million m^3)	2646	908	77.87	193.1	229	68.25
Maximum depth (m)	37.5	33	10.1	29.3	30.2	70.0
Mean depth (m)	17	11.5	2.3	8.1	11.4	5.2
Catchment area (km^2)	16 300	4 200	–	2 253	10 826	5 428.43
Elevation (m)	243	280	38	279	222	483
Length of shoreline (km)	–	125	–	–	–	–
Shore dev. index	6.7	4.0	–	–	6.2	1.5
Volume dev. index	1.4	1.0	0.69	0.8	1.0	0.94
Annual level fluctuations (m)	–	22.6	–	–	–	–
Inflowing rivers	Cauvery	Bhavani Moyar	Kortaliyar Nagari	Vaigai	Ponnair	Ponnair
Outflowing rivers	Cauvery	Bhavani	–	Vaigai	Ponnair	–
Latitude (N)	11°49’	11°30’	–	10°01’	12°12’	10°30’
			Soil			
pH	–	5.2–6.4	–	–	–	–
Organic carbon (per cent)	–	1.83–2.68	–	–	–	–
Total nitrogen (per cent)	–	0.14–0.26	–	–	–	–
Available nitrogen (mg l00 g^{-1})	–	31.0–50.8	–	–	–	–
Available phosphorus (mg l00 g^{-1})	–	1.0–3.58	–	–	–	–
			Water			
Water temp. (°C)	24.2–32.0	25–57	27.0–31.6	–	25.0–30.4	31.4–23.5
pH	7.5–8.8	7.8	8.8–9.6	7.5–8.5	8.3	8.0
Transparency (cm)	–	18.5–135.0	66–55	–	–	11.5–70
DO (mg l^{-1})	4.5–12.6	7–48	4.8	–	8–5	6–8
CO_2 (mg l^{-1})	–	3–90	0–2.0	trace	–	–
Total alkalinity	140–278	40–92	229–554	79–247	145–616	146–373

Parameters	*Stanley*	*Bhavanisagar*	*Poondi*	*Vaigai*	*Sathanur*	*Krishnagiri*
Spec. cond. (umhos)	170–350	269–96	500–400	125–300	320–800	425–575
Total hardness (mg l^{-1})	86–128	26.11	98–66	65–280	112–254	76–146
Calcium (mg l^{-1})	tr.	–	33.5–11	16.5–17.5	–	–
Nitrates (mg l^{-1})	tr.	tr.–0.34	Nil	–	4.4–5.6	–
Phosphates (mg l^{-1})	0–0.1	0.01–0.06	Tr	–	–	–
Silcates (mg l^{-1})	0–19.9	0–14.3	22.8–5	0–9.4	0.1–32	0.34–4.4
Chlorides (mg l^{-1})	19–37	9–30	23–88	10–23	18.66	35–120
Organic carbon (mg l^{-1})	1.8–12	6.6–9.0	–	–	7.8–12.6	1.26–8.58
Level fluctations (m)	–	5.5–22.6	–	–	–	–

Parameters	*Willingdon*	*Pechipparai*	*Manimuthar*	*Amaravathi*	*Vidur*	*Aliyar*
Year of sealing	–	–	1958	1958	–	1962
Area (ha) at FRL	1 554	1 515	940	850	798	646
Volume (m m^3)	573.4	126.4	146	112	16.93	107
Maximum depth (m)	–	37.7	36.0	33.8	9.8	36.5
Mean depth (m)	3.86	8.4	16.8	13.7	2.1	18
Catchment area (km^2)	129.5	207.19	161.61	832	1298	468.80
Elevation (m)	72.0	92	110	358	38	320
Length of shoreline (km)	–	–	–	–	–	16
Shore dev. Index	–	–	–	2.3	–	–
Volume dev. Index	–	0.75	1.4	1.2	0.68	1.2
Annual level fluctuations (m)	–	–	–	–	–	31
Inflowing rivers	Periya Odai	Kodayar	Manimuthar	Amaravathy	Varahanadhi	Aliyar
Outflowing rivers	–	Kodayar	Manimuthar	Amaravathy	Varahanadhi	Aliyar
Latitude (N)	–	–	–	10°15'	12°04'	10°15'
Longitude (E)	–	–	–	77°15'	79°34'	76°50'
Soil						
pH	–	–	–	–	–	6.42
Sand (per cent)	–	–	–	–	–	24.1
Silt (per cent)	–	–	–	–	–	29
Clay (per cent)	–	–	–	–	–	46.9
Water						
Water temp (°C)	20.3–24.5	–	–	24–28.8	–	21.2–32.5
pH	8.2–8.4	6.6–7.2	6.9–7.9	6.7–9.1	8.7–8.6	6.6–8.8
Transparency (cm)	20–51	–	–	–	–	108–182
DO (mg l^{-1})	6.4–8.6	–	–	8.1	–	4.2–11.6
CO_2 (mg l^{-1})	0	–	–	–	–	nil–10

Parameters	*Willingdon*	*Pechipparai*	*Manimuthar*	*Amaravathi*	*Vidur*	*Aliyar*
Total alkalinity	120–150	28–24	12.5–28	7.84	840–570	16.0–58.0
Spec.cond. (μmhos)	240–320	32–19	29	38–63	80–150	38.7–109.6
Total hardness (mg l^{-1})	–	10	20–14	18–50	56–152	14–40
Calcium (mg l^{-1})	–	–	–	–	–	4.0–18.0
Magnesium (mg l^{-1})	–	5	9–16.7	–	14.5–26.5	1.2–6.2
Nitrates (mg l^{-1})	–	–	–	1.7–2.8	–	tr.0.04
Phosphates (mg l^{-1})	–	–	–	0.0–0.01	–	tr.0.02
Silicates (mg l^{-1})	6–10	1.8–3.0	1.6–5.7	1.4–38.5	1–26.5	2.8–24.0
Chlorides (mg l^{-1})	18–22	10–14	14–6	0.4–1.0	104–16	4–24
Organic carbon (mg l^{-1})	–	–	–	12.5–21.6	–	–
GPP ($gO_2m^{-2}d^{-1}$)	–	–	–	0.3–18.3	–	0.278–0.699

Parameters	*Hope Lake*	*Pykara*	*Uppar*	*Tirumoorthy*	*Gomukhi*	*Sholiar*
Area (ha) at FRL	573	448	453	388	360	553
Volume (m m^3)	216.2	59.69	15	51.0	15.86	147.2
Maximum depth (m)	55	36.6	–	32	14	–
Mean depth (m)	37.7	14.5	–	11-	4.6	27.4
Catchment area (km^2)	57.5	97.5	903.88	80.29	292.67	120
Elevation (m)	244	2036	276	407	183	370
Shore development index	–	–	–	–	2.6	–
Volume development index	2.0	1.2	–	1.0	1.0	1.01
Inflowing rivers	–	–	Uppar	Palar	Gomukhinadi	Sholiar
Latitude (N)	–	–	10°47'	–	–	10°10'
Longitude (E)	–	–	77°25'	–	–	76°55'
Water Temp. (°C)	–	–	22.5–29	–	25.1	22.6–24.2
pH	6.6–8.6	6.9	8.0–8.8	7.7	7.8–8.6	6.7–7.1
Transparency (cm)	–	0.77–3.78	–	–	25.5	–
DO (mg l^{-1})	–	–	4–10	7.6	5.7	8.0–8.2
CO2(ml l^{-1})	–	–	Nil	–	–	2.2–4.0
Total alkalinity (ml l^{-1})	10–21.3	6.0–18.0	56–182	16–124	88.6–309	22.0–32.0
Spec. cond. (umhos)	30	14–45	–	40–200	230–550	40–70
Total hardness (mg l^{-1})	8.0	10.4–18.0	–	16–44	68–124	20.0–28
Calcium (mg l^{-1})	4	Trace	–	–	24	–
Phosphates (mg l^{-1})	–	1	–	0–2.0	Trace	nil-0.06
Silicates (mg l^{-1})	1.4–3.0	–	–	5–16	19.3–6.1	6
Chlorides (mg l^{-1})	1.0–18.0	10.0–18.2	–	2–14	24.0–54	2–12
Organic carbon (mg l^{-1})	–	–	–	4–14.5	–	8.5–10.2

Parameters	Sandynulla	Pambar	Manimukta	Goddar	Pilloor
Area (ha) at FRL	258	243	746	678	389
Volume (m m^3)	20.28	7	20.6	153.60	44.4
Maximum depth (m)	24.7	–	–	–	–
Mean depth (m)	10.3	14	10.97	8	11.4
Catchment area (km^2)	43.5	1736	484.31	–	117
Elevation (m)	2120	321	128	–	920
Shore dev. index	4.6	–	–	–	–
Volume dev. index	1.2	–	–	–	0.46
Inflowing rivers	–	Pambar	Manimukta	–	–
Latitude (N)	–	–	–	–	11° 16'
Longitude (E)	–	–	–	–	76° 49'
Water Temp (°C)	22.6–16.4	23.2–29	23	19–28	26–29.6
pH	7.95	8	8.4	8.3–8.7	7.9–8.6
Transpareny (cm)	78.5	–	32.5	40–70	–
DO (mg l^{-1})	8.3	7.2	7.4	4.4–7	7.6–8.8
Co_2 (mg l^{-1})	2.4	NIL	5	0	nil—1.4
Total alkalinity	26–62	396—398	153	274–350	36.5–85.0
Spec. cond. (μmhos)	99–280	300–950	315	550–680	60–125
Total hardness (mg l^{-1})	28–74	204–208	–	–	28
Calcium (mg l^{-1})	–	–	–	–	7–11
Nitrates (mg l^{-1})	1.5–2.1	–	–	–	–
Phosphates (mg l^{-1})	0-Trace	tr–0.2	0.02	–	nil
Silicates (mg l^{-1})	0–4.7	10	34	4–6	6.2–20.8
Chlorides (mg l^{-1})	12–48	–	15	30–40	4–12
Organic carbon (mg l^{-1})	10.05	–	–	–	4.7
GPP (g O_2 m^{-2} d^{-1})	6.1	–	–	–	–

Post Harvest Arrangements

Most of the ice plants and cold storages in the State cater to the marine sector. Mettur, Bhavanisagar, Sathanur, Krishnagiri and Amaravathy reservoirs have ice plants and cold storages for storing fish before its despatch to the Howrah railway station in Calcutta. Various market channels are in vogue in respect of reservoir fishes. The agencies involved are primary cooperative societies, Tamil Nadu Fisheries Development Corporation (TNFDC), Department of Fisheries and the private traders. Networks involving one or more of the above agencies are also not uncommon. For instance, the TNFDC collects fish from Aliyar, Sathanur, Bhavanisagar and Tirumoorthy and sells it directly to the public or to merchants. There are also arrangements between the Department of Fisheries and TNFDC to dispose the catch in the event of glut. A substantial part of the catch from Sathanur, Mettur, Bhavanisagar and Aliyar is despatched to the Howrah market.

Cooperatives

Tamil Nadu has 598 primary fishermen's cooperatives with a total membership of 157 400, of which 250 are inland fishermen societies. These societies are not as efficient as their counterparts in agriculture, marketing, consumer and housing sectors, primarily because the fishermen cooperatives confine themselves mainly to the provision of cheap credit (long, medium and short-term) and working capital to fishermen (Parasuraman, 1979). They do not attract institutional finance due to legal restrictions. Fishermen cooperatives in the State are financed solely by the State Governmentin the form of share capital and loans in the long, medium and short-term categories. They normally do the following functions:

1. Distribution of essential commodities to their members with the working capital accommodations granted by State Government.
2. Taking small reservoirs on lease and let them on sub-lease to members on yearly basis.
3. Fish marketing, and
4. Distribution of yarn under subsidy schemes.

Chapter 6

Recent Advancements in Reservoir Fisheries Management

'Aquatic ecosystems' including ponds, rivers, lakes, reservoirs, flood plains, coastal lagoons, estuaries, coastal shelves and open oceans cover a very large part of the earth's surface. Human interferences like overfishing, damming, pollution, habitat destructions, introduction of exotic species, *etc.* are totally upsetting most of the aquatic ecosystems. As it is difficult in managing single species of fish or any other aquatic organism in a large diverse aquatic ecosystem, fisheries management is gradually migrating from a single species approach to ecosystem based fisheries management approach. The ecosystem based fisheries management need a comprehensive knowledge about the structure and functioning of the aquatic ecosystem. That knowledge provides a basis for envisaging how natural distresses and human activities will influence the delivery of desired ecosystem goods or services and hence form basis for a scientifically sound management.

Modelling is an essential scientific tool in developing ecosystem approaches for fishery management and there occurred so many developments in building of trophic models of aquatic ecosystem. Ecosystem models play an important role in supporting ecosystem approaches for sustainable utilisation of resources by studying the aquatic ecosystem as a whole. Sustainable use of living aquatic resources takes into account, both the impact of ecosystem on the resources and fishery or other activities on the ecosystem. Thus,a modelling approach which helps to estimate the trophic level of the species or functional groups existing in the system is essential to understand the trophic interactions of aquatic ecosystem.

Ecopath, a trophic modelling approach, has been used to model a wide variety of aquatic ecosystems. Polovina initiated the Ecopath approach in 1984 and has been under continuous development since 1990. Ecopath is the core routine of the Ecopath with Ecosim (EwE) software, which is a widely-used tool for analysis of exploited aquatic ecosystems, as it helps to describe the complex trophic relationships in aquatic ecosystems on quantitative basis.Ecopath is used to establish mass balance which allows analysis of flows between trophic levels and there by helps to study the status of the ecosystem in general.Hence, Ecopath can be stated as a potential predictor of the gross impacts, of large scale exploitations happened on an ecosystem.

Ecopath is a static mass balance model which once established in an ecosystem, helps to know about the available resources in the system and also provides information on interactions between various resources.Therefore, Ecopath is not a simulation model, unless it describes ecosystem at steady state for a given period. The species or functional group in the ecosystem has to be specified by the user for describing the ecosystem. The Ecopath model analysed the maturity of the system as a whole. An undisturbed aquatic ecosystem is considered to be mature and the disturbances to the system, notably by fishing, reduce the maturity of the system. The further development of the Ecopath model leads to origin of dynamic ecosystem model called 'Ecosim', which is capable of simulating ecosystem changes over time. Ecosim simulations were conducted to examine changes in ecosystem structure and maturity under different management scenarios. Another component is 'Ecospace' which is a spatial and temporal dynamic module primarily designed for exploring impact and placement of protected areas. The present article was meant to give a brief knowledge about Ecopath model, its data requirements and the final estimations about the ecosystem that can be developed by using this model.

6.1. Methodology

Ecopath with Ecosim Software

Ecopath with Ecosim (EwE) is a free ecological/ecosystem modelling software suite. The software was designed for the creation of data-rich ecosystem models which best suits the ecosystem-based fishery management. Hundreds of ecosystem models have already been developed in various ecosystems and it can offer a better awareness into system evolution over time in response to changes in productivity and exploitation. Ecopath helps to analyse energy flow between species (or groups of species) in an aquatic ecosystem based on biomass estimates and food consumption relationships. Ecopath with Ecosim(EwE) modelling software is a commonly used model for exploring fisheries management questions within an aquatic ecosystem.

The Ecopath software package can be used for developing management policy options, to study the impact of fishing on ecosystem, impact study of marine protected areas, analysing environmental changes and to facilitate end-to-end model construction. The Ecopath model is based on the assumption of mass balance and based on two master equations, one to describe the production term and one for the energy balance for each group.

Production= Fishing mortality + predation mortality + migration + biomass accumulation + other mortality

Consumption= Production + Unassimilated food + Respiration

Ecological Groupings

The aggregation of the species into functional groups is essential to perform the network analysis of natural ecosystem. Hence while using Ecopath; the ecosystem is partitioned into boxes (groups) comprising species having a common physical habitat, similar diet and life history characteristics. Ecological groupings are made taking into consideration that; within group, the species have similar sizes, similar population parameters, similar food and similar predators.

6.2. Basic Input Parameters

The input parameter collection is major hurdle in developing the model (29).Necessary model inputs for each group include biomass (B) in tonnes/km^2, production/biomass ratio (P/B) per year, consumption/biomass ratio (Q/B) per year, ecotrophic efficiency (EE), diet composition, and any associated fishing mortality per year. Out of the four parameters (B, P/B, Q/B and EE), usually EE is difficult to estimate and be estimated by the model if the reasonable estimate of other three parameters are given.

The average biomass estimates for functional groups can be derived from the primary literature, experimental fish survey data or stock assessments, or can be estimated based on commercial catch data and fishing mortality rates. For exploited groups, average biomass of each group per unit area can be estimated from the equation of Gulland,

B=Y/F (16)

where Y is the annual average yield and F is the Fishing mortality (Z-M). The biomass for unexploited groups like phytoplankton, zooplankton, zoobenthos, dipterans, macrophytes, benthic algae, fish eating birds *etc.* can be obtained from field studies and also from similar ecosystems

The P/B ratio is equivalent to the instantaneous rate of total mortality (Z) commonly used in fisheries modelling. For fish group, the total mortality can be calculated from the length converted catch curve equation or calculated as a sum of natural mortality and fishing mortality. The natural mortality can be estimated using the Pauly's empirical formula. For unexploited population the total mortality is equal to the natural mortality. The value can also be collected from the literature related to similar environment. The consumption/Biomass (Q/B) ratio for the predator groups can be either calculated using empirical regression formula or taken from other models for the similar ecosystem. This parameter expresses food consumption (Q) per unit biomass for a conventional period of one year, which means number of times a given population consumes its own weight per year. The empirical equation for the calculation of Q/B relates mean annual temperature, body size, and morphometric aspects of the body and the caudal fin.

Ecotrophic efficiency (EE) is the proportion of the ecological production which is consumed by the predators and/or exported (Ricker, 1969) and is difficult to estimate.

Collection of Diet Composition Data

The model also requires the input of dietary composition of each group. Since the food web links the different functional groups in an ecosystem, information on diet composition is important for understanding the dynamics of the ecosystems. The study of the diet content is also helpful in developing the prey-predatory relationship. Diet composition of consumers will be represented in a prey- predator matrix that contains the fraction of the diet of each predator contributed by each prey. The diet composition of all consumers must be entered, if absent it can also be collected from literatures about same species in the similar environment.

Balancing the Model

The model has to be balanced by adjusting the basic input parameters until all the groups obtain the ecotrophic efficiency less than one. This can be done as the ecotrophic efficiency values greater than one are not plausible *i.e.* it is not possible that more biomass is used than produced by a group under the steady state condition.

Ecosystem Indicators

The ecosystem indicators which helps to explain the final results obtained using Ecopath model are listed below

System Omnivory Index (SOI)

The omnivory index (OI) indicates that species ranged from highly specialised consumers to generalist predators. A higher value of OI shows that these organisms feed in a diverse range of trophic levels. The OI values range from 0 to 1, where '0' indicate a highly specialised consumer and '1' indicates a generalised consumer. The SOI is the average OI of all consumers weighted by the logarithm of each consumer's food intake. The SOI is a measure of how the feeding interactions are distributed between trophic levels. The system omnivory index therefore characterizes the diversity of consumer–prey relationships.

Total Primary Production/Total Respiration

In a mature system, the ratio between total primary productivity and total system respiration (TPP/TR) would approach unity, whereas smaller ratios are characteristics of system under developmental stage. In a system under early developmental stage the rate of primary production exceeds rate of community respiration, hence the ratio will be higher. The ratio is considered to be an important ratio for the description of maturity of an ecosystem.

6.3. Net System Production

Net system production or yield is the difference between total primary production and total respiration. System production will be large in immature system and close to zero in mature ones. The unit will be t Km^2/year.

6.4. Total Primary Production/Total Biomass

The between system's primary production and its total biomass is expected to be a function of its maturity and it take any positive values. In immature systems, production exceeds respiration and the ratio take higher values, while in a mature system the value of the ratio becomes lower.

6.5. Total Biomass/Total Throughput

The total system biomass is supported by the available energy flow in the system and it can be expected to increase to a maximum for the mature stage of a system.

Connectance Index

The CI for a given food web, can be defined as the ratio of the number of actual links to the number of possible links, where feeding on detritus (by detritivores) is included in the count, but the opposite links (*i.e.* detritus 'feeding' on other groups) are disregarded. Number of possible links in Ecopath can be estimated as (N-1) It has been observed that the actual link in a food web is roughly proportional to the number of groups in the system.

Electivity Index

The selection index or electivity describes a predator's preference for prey. The value ranges from '-1' to '+1'; where '-1' indicates the total avoidance of prey, '0' indicates that a prey is taken in proportion to its abundance in the ecosystem and '+1' indicates total preference for a prey.

Ascendency

Ascendency is seen as a measure of ecosystem growth and development. It is a key index that characterise the level of development and maturity of a system. The higher value of ascendency can be related to the low maturity of the system. The difference between the capacity and the ascendency is called 'system overhead'. The overheads provide limits on how much the ascendency can increase and reflect the system's 'strength in reserve' from which it can draw to meet unexpected perturbations.

Cycling Index

The cycling index is a fraction of ecosystems throughput that is recycled. Finn's Cycling index is an indicator of system stability and resilience. It measures the proportion of the total system throughput that is recycled and mature systems have higher recycling rates and recover faster from perturbations.

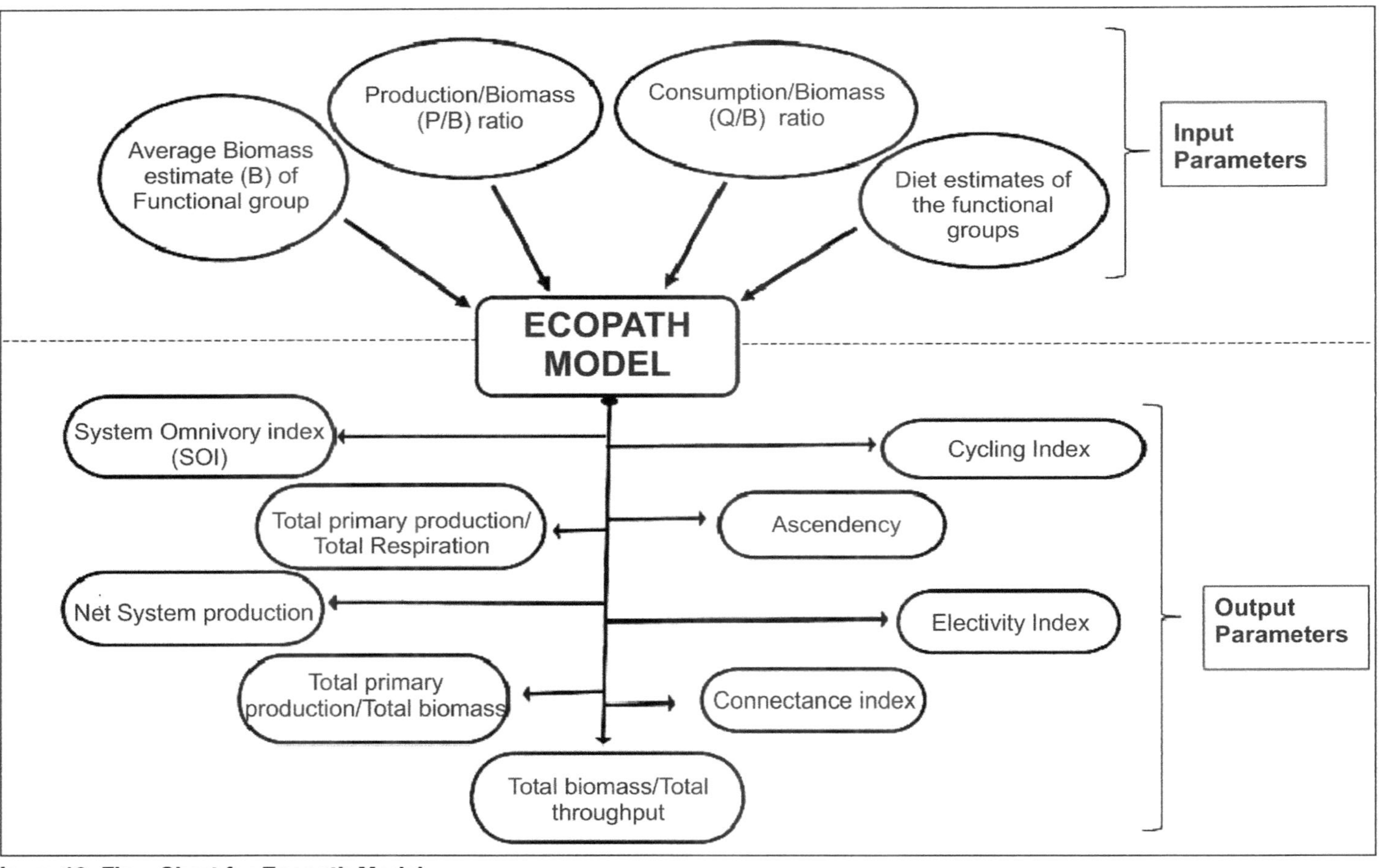

Figure 19: Flow Chart for Ecopath Model.

6.6 Socio-economic and Institutional Setting

In India, very few studies have been carried out on the socio-economic status and institutional framework of reservoirs, particularly with reference to fishing activities. In most reservoirs, fisheries have been developed as secondary user so that the fisheries component was missing during the planning phase of reservoir construction. As a result, there were no fisheries-related studies during the pre-impoundment phase. In most reservoirs, impediments to fishing such as submerged tree stumps, rocks, *etc.* are present and their removal after the construction of reservoir is costly. Representation from fisheries authorities in the reservoir management boards is therefore important for better fisheries management of the reservoirs. These water bodies, if managed properly for fisheries operations, have been proven to be very effective in enhancing income of rural communities in addition to providing employment opportunities as in the case of Gobindsagar and Pong Dam reservoirs in Himachal Pradesh (Katiha, 1994).

Since reservoir fisheries are generally considered a secondary activity, compared to the irrigation and hydro-power generation, the department of fisheries or other fisheries agencies generally obtain the fishery management rights from the owners of the reservoirs, either by paying some royalty or nominal amount (Rihand in Uttar Pradesh and Kanspati in West Bengal) or free (Gobindsagar and Pong Dam in Himachal Pradesh). In some situations, the Fisheries Department transfers the fishing rights to some other government agencies, co-operatives or private agencies and receives the royalty with or without rendering any fisheries development services.

One of the most important institutional arrangements for exploitation of fishery resources in reservoirs is a leasing system. It depends on trends of fish production, income and expenditure of the department, socio-economic conditions of the fishers, income and expenditure of the department, socio-economic conditions of the fishers, government policy towards co-operatives and status of fisheries resources of the state. In Uttar Pradesh and Rajasthan, the reservoirs with area i) below 100 ha are leased every year, ii) 100-500 ha for 3 years, iii) 500-1000 ha for 5 years, and iv) above 1000 ha for 10 years. In Madhya Pradesh, reservoirs below 100 ha are under the control of Gram Panchayat (*i.e.*, the local level administrative institution) for fisheries management, while reservoirs of 100-1000 ha are under Department of Fisheries and those above 1,000 ha are under the State Fisheries Corporation. Past experience has revealed that reservoirs entrusted to state agencies for management of fish stocks through stocking and control of harvesting have consistently maintained a good production level, while those under the control of private contractors have not been encouraging. Where the fisheries management was vested with cooperatives, the results have been moderate, but proper supervision at stocking and fish harvesting stages is required. When reviewing the fisheries management practices for various sized reservoirs, the appropriate management strategies for large and medium-sized reservoirs should involve stock enhancement aspects, while for the small reservoirs, culture based management is considered to be appropriate.

The unavailability of reliable catch statistics and yield estimates in reservoirs is a major setback for defining scientific management strategies for Indian reservoirs. More discomforting is the fact that the production figures available on most of the reservoirs are inaccurate and unreliable. This lacuna is attributed to several factors, chiefly:

- ☆ The multiplicity of agencies owning fishing rights creating difficulties in gathering data in some states.
- ☆ Highly scattered and unorganised market channels, mostly under the clutches of illegal money lenders.
- ☆ Ineffective cooperative set ups.
- ☆ Diverse licensing/royalty/crop sharing systems practiced by different state governments, some of which include a free-for-all system, providing little scope for recording catch statistics.
- ☆ Inadequate and poorly trained manpower at the disposal of state governments and cooperatives to college catch data, follow statistically sound sampling procedures and to cover the whole reservoir.

There is a need to merge the twin objectives of conservation and yield optimisation in reservoir fisheries management. While the fishers and the fish vendors strive to increase the production to maximise profit, it is the responsibility of the state to ensure that economic expediency of development does not mark ecological reasoning. The virtually open access nature of fisheries exploitation, as practiced in Andhra Pradesh, is counter-productive to resource conservation and management.

Although there are fair possibilities of linking reservoir fisheries development with poverty alleviation programmes, the progress made thus far in this direction is not overly encouraging. The chances of creating additional employment are not high in the majority of Indian reservoirs. On the contrary, many reservoirs have surplus manpower, which could be diverted to other reservoirs, enabling them to continue fishing without adverse effect on the income level of the existing fishers (Paul and Sugunan, 1990).

Proposed Action Plan

Measures to increase reservoir fisheries productivity include:

- ☆ Extensive transfer of technological backstopping to the State Department of Fisheries and grass root level fishers/fisher organisations.
- ☆ Re-appraisal of present state policies for the ownership, fishing rights and fisheries management, leasing (lessee, lease tenure and amount) to encourage the reservoir fisheries development activities.

- ☆ Enforcing appropriate guidelines for stocking and other management protocols, strict monitoring of stocking and harvesting schedules. This may include:
  - Issue notification for close season or fishing-holiday during the breeding season.
  - Declaration of the lotic sector of reservoir as a highly sensitive and protected area during the breeding season.
  - Effective enforcement of size and mesh regulations to conserve the breeding stocks.
  - Strict monitoring and prohibition of exotic species to build or conserve the stocks of economically viable and region specific fish species.
- ☆ A partnership approach may be encouraged by involving various user groups, such as irrigation, agriculture, fisheries and environment departments to provide enabling environment for sustainable fisheries, *e.g.* The irrigation authorities to assure minimum required level of water in reservoir for safe maintenance of biotic communities and information about opening of sluice gates to be shared with the fisheries department.
- ☆ Authorities to recognise fisheries as an important activity in reservoirs.
- ☆ Adoption of various options for stock and species enhancement for management of reservoir productivity including the pens and cages.
- ☆ Infrastructure support to provide desired quality and quantity of fish seed in time and space.
- ☆ Minimising losses by way of escape of fish through spillways and canals through designing suitable screen structures and their annual repairs.
- ☆ Introduction of community-based/co-management concepts considering the multiple use, multi-stakeholder environment in Indian reservoirs.
- ☆ Inclusion of fisheries experts in committees at the start of large water diversion projects and schemes.
  - Giving due importance for fishery activities at the pre-impoundment and/or planning stage of proposed reservoir.
- ☆ Providing infrastructure and credit support for marketing and post-harvest activities in reservoir fisheries.
- ☆ Development of mechanisms for long-term generation of a reliable database on reservoir fisheries including the morphometry, catch and fisher statistics, management and property regimes.
- ☆ Organising trainings and refresher courses for the personnel from state fisheries departments, reservoir managers, NGOs and other stakeholders to upgrade their knowledge on reservoir fisheries and ecological management.

☆ Viable schemes for financial assistance to fishers depending upon reservoir fisheries for their livelihoods.

Conclusion

The Ecopath model can be well utilized for the assessment of impacts of management measures on an aquatic ecosystem by showing the impacts on different functional groups before and after the management measures. It is already established as a powerful tool for the exploration of past and future impacts of fishing and other environmental disturbances on an ecosystem. The impacts of invasive aquatic organisms in the wild, which is a real issue faced in aquaculture as well, can be better assessed by developing a mass balance model using Ecopath with Ecosim software. Thus, Ecopath is really helpful in designing policies, aimed at implementing ecosystem management principles. The data of diet composition of all the consumers in the system are also essential for the development of the model. The model also has the provision estimation of the unknown basic input parameters which are unknown, if the better estimate of all other input parameters is available.The estimation of the ecosystem indicators helps to explain the maturity state of the system which is the prime need of using this model. The diverse interactions (positive or negative) among various functional groups in the system can be explained well from the results.The major problem in using the mass balance model like Ecopath is the availability and collection of the basic input information. The Ecopath model, a mass balance model is really helpful in developing strategies for the management of aquatic ecosystem through proper understanding about interactions among the different ecological components.

Chapter 7

Cage and Pen Culture in Reservoir

7.1. Cage Farming in an Indian Reservoir

Presently the annual fish yield from small, medium and large reservoirs is 100, 75 and 50 kg/ha, suggesting substantial scope to enhance fish yield through wild capture and culture-based fisheries in these water bodies. The success of auto-stocking is very low in Indian reservoirs, especially smaller ones, often a result of recruitment failure. Therefore, regular and sustained stocking is needed to augment fisheries in reservoirs. Stocking with the appropriate species, size (>100 mm) and at the right time is essential to optimize fish yield from reservoirs. Production of fingerlings in cages offers the opportunity to produce fingerlings that can be used in stock enhancement programs for Indian reservoirs. Apart from raising fingerlings for reservoir stocking, cages can also be used to support production of high-value table-fish and crustaceans.

History

Rearing of fish in cages was first described by Lafont and Saveun in 1951 (cited by Hickling, 1962). Cage culture possibly first originated nearly 200 years ago in Cambodia where fishermen used to keep *Clarias* sp. and some other fishes in bamboo cages in the basements of floating dwellings, primarily for holding and marketing. This gradually became a system of culture, since the fish could withstand crowded conditions in the cages and grow with kitchen scraps, left over rice and other similar waste materials and gained popularity throughout the lower Mekong basin of the country (Ling, 1977). Cage culture is traditional in parts of Indonesia also,where cages are anchored in streams,which are practically open sewers. Common carp culture in bamboo cages is practices in west Java, since the early 1940. This type of traditional fish culture,distinguished by its reliance on natural construction materials and waste feeds,is still practiced in many parts of Indionesia and Indo-China. However, although moderately successful,these methods of rearing fish have been largely localized and not directly given rise to the current cage fish farming industry. Modern cages utilize synthetic mesh or netting and have collars, largely fabricated from synthetic polymers and metals, although wood is still widely used in many designs.

Standing water cage culture probably,originated in Japan in the early 1950s which has been adopted for commercial culture of fishes in most countries including India. However, it was only during the past 20 years that aquaculture in enclosures has spread (now practiced in over 35 countries) and has become a high-tech business in the developed countries such as floating and/or submerged cage culture of salmonids in Canada, Norway and Scot land, tuna in Japan and catfish in Southern USA. The same trend may be expected in the tropics also, as large business houses began to take interest in this system of aquaculture.

In India, cage culture was attempted for the first time in the case of air breathing fishes in swamps which are marked by low dissolved oxygen in water and for major carps in running water in Yamuna and Ganga during 1972-76 at Allahabad and common carp, catla, silver carp, murrels and tilapia in a still water body in Bangalore. The cages also have been used for replacing ground nurseries for rearing fry.

7.2. Evolution of Freshwater Cage Culture in India

In India, cage culture was attempted for the first time in 1970 in three environments:

1. Swamps marked by low dissolved oxygen concentration, using air-breathing fishes.
2. Running waters of the Yamuna and Ganga Rivers at Allahabad, using major carps.
3. A static water body in Karnataka, using common carp, catla, silver carp, rohu, snakeheads and tilapia.

The Central Inland Fisheries Research Institute (CIFRI) attempted cage culture in the 1970s with the production of air-breathing fish in cages but got poor results.

A few preliminary trials on cage aquaculture were attempted and recently CIFE, Mumbai did some isolated work on cage culture for raising fingerlings as well as table-fish in reservoirs such as Powai (Maharashtra), Govindsagar (H.P.), Halali (M.P.), Tandula (Chattishgarh) and Dimbe (Maharashtra). Later, a number of attempts were made to produce cage-stocked fish, especially fry to fingerlings (Banerjee *et al.,* 1979). Culture of silver carp and common carps was tried in floating cages in Sankey Tank, Bangalore, but success was unremarkable. With the lessons learned, cages were subsequently used for raising catla fingerlings in Karnataka (Govind *et al.,* 1988), also with little success. Production of fingerlings for stocking reservoirs was tried in Govindsagar (H.P), Getalsud (Ranchi), Gularia (U.P.) reservoirs, with poor success because proper monitoring was not initiated.

Cage aquaculture was initiated in CIFA at Hessaraghata farm, Bangalore with one unit established in 1997-1998 and later at Kaushalyaganga farm, Odisha in a pond system with major improvements in cage design. While the objectives of cage aquaculture were not properly understood, these experiments served as good demonstrations.

More recently cage aquaculture experiments were conducted in Walvan Reservoir (Maharashtra), with fry to fingerling culture of *Tor putitora* and *T. khudree* (Kohli *et al.,* 2002). During 2005-2006, floating cage culture experiments were conducted in Kabini reservoir, Karnataka for raising fingerlings for stocking, with moderate success, defined by a survival rate of around 40 per cent. Some trials on cage aquaculture were conducted during 1998 onward for production of fingerlings but results were not optimal. CIFRI has banked on the success of producing fingerlings for stocking reservoirs in floating cages installed in different agroclimatic zones of India (Madhya Pradesh, Uttar Pradesh, Bihar, Jharkhand and North-east states).

What is Cage Culture?

Cage culture is when fish are reared from fry to fingerling, fingerling to table size, or table size to marketable size while captive in an enclosed space that maintains the free exchange of water with the surrounding water body. A cage is enclosed on all sides with mesh netting made from synthetic material that can resist decomposition in water for along period of time and is sold under the brand name Netlon. Cages are generally small, ranging in freshwater reservoirs from square meter (m) to 500m. Several small cages combined in a battery, as described below, are suited for even intensive culture.

Why Cage Culture?

The reservoirs of India have a combined surface area of 3.5million hectares (ha), mostly in the tropical zone, which makes them the country's most important inland water resource, with huge untapped potential. Fish yields of 50kg/ha/year from small reservoirs, 20kg/ha/year from medium-sized reservoirs and 8kg/ha/year from large reservoirs have been realized while still leaving scope for enhancing fish yield through capture fisheries, including culture-based fisheries. The success rate of auto stocking is very low in Indian reservoirs, especially in smaller ones. Many of the smaller reservoirs dry up during the summer, partly or completely, with no stock surviving. A policy of regular, sound and sustained stocking would greatly augment fisheries in such water bodies. The prime objective of cage culture discussed here is to rear a fingerlings measuring >100 millimetres(mm) in length, especially carp, for stocking reservoirs.

Stocking with the right fish species, using seed of appropriate size and introducing it at the right time are essential to optimizing fish yield from reservoirs.

Though billion fish fry are produced every year in India, there is an acute shortage of fish fingerlings available for stocking reservoirs. Where fingerings are available, transporting them to reservoirs usually incurs high fingerling mortality. In this context, producing fingerlings *insitu* in cages offers opportunity for supplying stocking materials, which are vital inputs towards a programme of enhancing fish production from Indian reservoirs.

7.3. Advantages of Cage Culture

Cage culture is suitable to a wide range of open freshwater ecosystems, especially reservoirs. It efficiently exploits water bodies, tapping their natural productivity and thereby reducing pressure on other resources. It uses simple technology and locally available resources for cage construction and operation, making it economically, socially and environmentally sound. As carp feed at a low trophic level, rearing carp fingerlings has minimal impact on the environment. Polyculture of carp species with various feeding habits makes wise use of resources, as the different feeding habits of various species and their acceptance of a wide range of supplemental feeds maximizes fingerling uptake of feed while minimizing competition among species, feed waste and the resulting pollution.

Table 29: Cage Materials and their Particulars

Name of Cage Material	*Particulars for Raising Fry to Fingerling*	*Particulars for Raising Fingerlings to Table Size*
Bamboos	One battery of 8 cage with cage size 5 m x 3 m x 3 m	One battery of 16 cages with cage size 4 m x 4 m x 3 m
Galvanized	2 inches dia x 3.6 mm thickness	2 inches dia x 3.6 mm thickness, 60 feet, 20 nos.
Iron Floats	24 metal drums, 22 kg weight, 88 cm length 180 cm circumference, 58 cm dia	64 metal drums, 22 kg weight, 88 cm length 180 cm circumference, 58 cm dia
PVC pipe		0.5 inches dia, 2.5 mm thickness, 260 feet
Iron brackets		130 x 60 cm long and 5 cm dia are required to tie the drums on both sides
Nut and bolts	Galvanized iron, 120 x 18 cm long and 3 cm dia	Galvanized iron, 100 x 18 cm long and 3 cm dia
Sinkers	48 stones of 3-4 kg 48 stones of 3-4 kg	48 stones of 3-4 kg 48 stones of 3-4 kg
Anchors	Large stones >40-50 kg	Large stones >40-50 kg
Net	HDPE, 1.5 mm mesh size	HDPE, 15-24 mm mesh

Cage culture eliminates losses to predation and facilitates prophylactic measures to contain any outbreak of disease, allowing very high fingerling survival rates. It makes effective use of manpower, as daily maintenance routines and monitoring are relatively simple, and harvesting is rapid, easy, sure and complete. As cage culture can be practiced intensively, high yields can be achieved very cost effectively. Since most reservoirs in India are designated for multiple uses, including supplying drinking water, cage culture is appropriate because it is minimally polluting and maintains the ecological health of the reservoir.

Table 30: Fingerlings to be Selected for Cage Rearing to Table Size

Species	*Number of Cages*	*pcs/m³*	*Total Number (including transport mortality)*
L. rohita	2	25-30	2000
C. carpio	2	25-30	2000
M. rosenbergii	2	15-20	1500
A. testudineus	2	50-70	5000
C. batrachus	2	30-40	3000
P. sutchi	5	30-35	6000
L. bata	1	50-100	2500
TOTAL			22,000

7.4. Species Selection for Cage Culture

Species selection depends on local interest and market value. For fingerling production, the species selected were primarily carp. During winter, fish had access to natural food sufficient for further growth after stocking in the reservoir. For growing table-fish, the species must have high market value, such as freshwater prawns, air-breathing species, seabass and, to some extent, carps and Pangasius sp. Small indigenous fish species (SIFS) with high market demand are also suitable.

Local areas must be surveyed to assess the availability of seed and, if not available, they can be transported from West Bengal, where seed is available and costs less than any other part of India. Fry or fingerlings packed with pure oxygen can be transported for the 40-h journey from West Bengal. Because different size groups prevail among fry, they should be graded and a size of around 80 mm selected for stocking if table-fish culture is the objective.

Cage culture makes it easy to harvest fingerlings for direct stocking in reservoirs, thereby saving transport costs and losses from transport mortality. On average, at least 70 per cent survival of fingerlings can be achieved following the procedures outlined. In the case of table-fish culture, 70-80 per cent survival is assured (Das *et al.,* 2009).

7.5. Steps of Cage Culture

1. Types of cage
2. Site selection
3. Procurement of cage materials
4. Frame fabrication
5. Floating the frame

6. Fitting the catwalk
7. Installation of cages
8. Selection of stocking materials
9. Stocking
10. Grow-out period
11. Supplementary Feeding
12. Cage and stock maintenance
13. Removal and release
14. Economics of cage culture
15. Cost of production of each crop

7.6. Types of Cage

Four types of cage are used in cage aquaculture: fixed, floating, submersible and submerged. The fixed cage is the most basic and widely used in shallow water with a depth of 1-3 metres. It consists of net bag fitted to posts and is normally placed in the flow of streams, canals, rivers, rivulets, shallow lakes and reservoirs, not touching the bottom. Fixed cages are comparatively inexpensive and simple, but their use is restricted. Floating cages, on the other hand, are supported by a floating frame such that the net bags hang in water without touching the bottom. Floating cages are generally used in water bodies with a depth of more than 5 metres. Enormous diversity in size, shape and design has been developed for floating cages to suit the wide range of conditions of fish culture in open waters. The net bags of submersible cages are suspended from the surface, have adjustable buoyancy, and may be rigid or flexible. Submerged net bags are fitted in a solid and rugged frame and submerged under the water. Their use is very limited.

7.7. Site Selection

The selection of site for cage culture is very important, as success often depends largely on proper site selection. Potential sites vary according to the size and shape of the reservoirs where cages are to be installed. The critical issues in selecting sites are the following:

1. The depth of the water column should be at least 5 metres. Water quality and circulation should be good, free from local and industrial pollution.
2. In large and medium-sized reservoirs, sites should be in sheltered bays for protection from strong winds. In small reservoirs, the cage should be anchored in the deeper lentic sector to avoid the current flow through sluice gates and irrigation channel.
3. They should be safe from frequent disturbance from local people and grazing animals.
4. There should be access to land and water transportation.

5. They should be devoid of algal blooms to avoid fouling.
6. They should be free of aquatic macrophytes and high populations of wild fish, which can cause oxygen stress.
7. Cages should be placed where they will not hinder navigation.
8. They should be at a distance from bathing and burning ghats.
9. Sites should be secure.

7.8. Procurement of Cage Materials

Making cage culture economically viable demands the preparation of a comprehensive list of materials available on local markets:

Bamboo

Bamboo poles should be straight, rigid and light, such as the bhaluka type from Assam, which is commonly available in local markets. Poles should be 7.5 metres long, with an internode circumference of about 26 centimetres (cm) at the base, 25 cm in the middle and 24 cm at the tip. This means a diameter of 8-9 cm and wall thickness of 2.5 cm at the base. Fifty-six bamboo poles are required to make one frame.

Floats

Empty 200-liter steel drums with tightened lids make suitable floats 88 cm long, 180 cm in circumference, 58 cm in diameter and weighing 22 kg. Twenty-four drums are required to float a frame 13.75 metres long and 11.05 metres wide, with drums sandwiched between upper and lower frames Floats are painted with acrylic paint and fastened with the frame with glazed iron (GI) wires.

Nuts and Bolts

Steel bolts 18 cm long and 3 cm in diameter, with nuts, are used to fix bamboo poles at the corners and at the middle joints on the side. A total of 120 such nuts and bolts is required to ensure a sturdy bamboo frame able to withstand waves even during a cyclone.

Sinkers

Sinkers are locally available stones weighing 3-4 kg that are tethered to the bamboo frame with nylon ropes at the corners and along the sides to help cages maintain their rectangular shape. The bottom portion of Netlon cages is tied to the nylon ropes running down to the sinkers such that the Netlon does not bear the weight of the sinker but the ropes maintain the shape of the cage. Eight Netlon cages require 48 sinkers, with eight sinkers maintaining the underwater shape of each cage.

Anchors

Large, locally available stones weighing 40-50 kg or more are used as anchors that rest on the reservoir bottom to hold the cages in place. Two anchors are tied with Garware-brand thick nylon rope to every corner of the frame, and another anchor is tied to the centre of each long side, thus requiring 10 stones for each cage.

Netlon

High-density propylene extract (HDPE) plasto nets with 1.5 mm mesh are used to prepare Netlon cages for rearing fingerlings to >100 mm from fry initially measuring 10-25 mm. A rectangular cage measuring 5x3x3 metres is convenient to operate. The cage is totally enclosed with Netlon on all four sides, the bottom and the top (to prevent predation by birds). Small flap openings at two top corners allow feeding and harvesting. The Netlon should be well stitched with doubly laced nylon ribbon 3.8 cm in width at the corners and joint, with loops at the corners and sides. The same nylon ribbon stitches together the upper and lower lids at regular intervals to make the cage more sturdy. Eight cages are hung from bamboo frame and tied with the sinkers at the bottom corners to keep them straight and hanging vertical. Thus in a battery of eight net cages has a good functional volume of 360 m^3, of which 320 m^3 (40 m^3 in each cage) is under water. The net cages are tied with silk rope to the frame to keep them straight.

Frame Fabrication

The frame of the cage can be made from locally available bamboo, which is a cheaper option than wood, steel or polyvinyl chloride (PVC) and will last for at least 3 years, with 5-10 per cent of the poles replaced as needed. Two frames are required, one above water and the other below, to hold the floats firmly. The useful life of bamboo poles in the underwater frame is much longer than for those on top.

Fully grown, cured bamboo poles at least 7.5 metres long and 8-9 cm in diameter at the base are best suited to make the frame. To make a battery holding eight cages, each measuring 15 m^2, the battery should be 13.75 metres long and 11.05 metres wide. To make such a frame, 32 bamboo poles are required for the top frame and 24 for the lower frame.

Four poles are used to make one long side of the rectangle. Two poles are placed on the ground with their tip portions overlapping to make their combined length at least 14.5 metres. Six such lengths are made for the two long sides and the central divider. Each long side and the central divider consists of two such 14.5-metre lengths placed parallel 35 cm apart.The short sides of the frame are made using similarly paired poles with a combined length of 11.75 metres. Ten such lengths are required for the two short sides and three partitions, again with two lengths placed parallel 35 cm apart. (The difference between the frame above the surface and the underwater frame is that the latter has only two partitions, instead of three, which is why it requires eight fewer poles.) Where the lengths cross at 90 degrees they are fixed with a bolt and nut, creating the rectangular frame of the battery, with inner partitions from which to hang eight net cages.

Floating the Frame

The battery of eight cages is buoyed by 24 steel drums. The drums are sandwiched between the two frames, one above the surface and the other below, placed mostly in the corners and near joints to provide the frame with balanced buoyancy. The drums are attached to the frame with two types of GI wire. The thicker, 2 mm wire is used at the central portion, and the thinner, 1.5 mm wire is used for encircling the drum with the frame at the end portions to better tolerate wave turbulence, especially during summer and storms. The drums are painted with acrylic paint before being attached to the frame. Only 16 cm of the drums' 58 cm of diameter is under the surface of the water when bearing the frame

Fitting the Catwalk

Catwalks made of locally available bamboo cross-beamed with wood are wired to the top of the bamboo frame with GI wire for ease of access. Care should be taken while wiring on the catwalks so that the wire ends do not damage the net cages. With the floats and catwalks wired to the frame, the assembly is towed to the selected site and anchored.

Installation of Cages

Once the frame is anchored at the culture site, the next step is to tie on the Netlon cages, eight to a battery.Along the top, silk ropes are used to tie the nets to the bamboo frame firmly to prevent sagging. Sinkers are tied to the bottom corners and the sides of the Netlon cages to hold them vertical. The hanging net cages should remain at least 1-2 metres above the lake bottom to avoid damage caused by crabs and other bottom dwellers. Local fishers should be instructed not to tie their gillnets to the frame, as this may damage it. The net cages should be left in the water for at least a week before stocking to allow algae on grow on the netting. Curing the net thus reduces injury to fry.

Stocking

For raising fingerlings in cages in Indian reservoirs, healthy carp fry measuring 12-15 mm long, or even up to 25 mm, are best suited. Advanced fry longer than 35 mm should be avoided for cage culture to fingerling size, as they routinely are affected by fungal diseases such as Seprolegniosis if collected from nurseries that have eutrophied. Indian major carps are especially prone to fungal diseases. A stocking density of 250 carp fry measuring 12 -18 mm per cubic metre is best for cages installed in Indian reservoirs. Fry should be shifted late in the day or early in the evening to allowing conditioning at the site of procurement and acclimatization at the site of release in cages. Conditioning is required to transport the fry with empty stomachs, as the ammonia and carbon dioxide generated by fish waste may prove lethal to fry during transport. Fry acclimatization is essential at the site of release in cages to ensure a balanced environment, especially in terms of temperature. The oxygen packets transported with the fry (1,000 fry in 4 litres of water in a polythene packet 2/3 filled with oxygen) are kept inside cages for at least an hour before the fry are released. Prior to release, fry are subjected to some prophylactic measures to protect them from diseases and ecoto-parasites. They are dipped in a 5-6 per cent salt solution as well as potassium permanganate (5-8 per cent) for 1 to 2 minutes and then released into the cage water.

Supplemental Feeding

Feeding is essential for growing fry to fingerlings in captivity because natural food in many Indian reservoirs may be insufficient for growth even to only fingerlings. In general, rice polish and mustard oil cake in equal parts, with a blend of vitamins, amino acids and minerals at 0.01 per cent was applied as supplementary feed for carp fry. In general, the fine flaky powdered form of rice polish/wheat polish/maize polish and mustard oil cake were mixed together, blended with vitamins-amino acid-mineral mixtures, and spread over the water surface in cages twice daily (0900 h and 1700 h) at 4-5 per cent of body weight. Initially 3-4 kg feed a day is applied per cage.

In case of table-fish production, floating feeds of high protein content must be selected. Extruded feed with 30-38 per cent protein was required for growing table-fish of carps, air-breathing species and prawns. Feeds remained floating on the water surface for a time, then settled slowly, thus progressively favoring the feeding habits of surface feeders, such as catla, first, then the column feeder rohu and finally the bottom feeders mrigal, common carp, air-breathing species and prawns.

7.9. Cage Maintenance

Regular monitoring of some water quality parameters, including dissolved oxygen, pH and free ammonia, inside cages is necessary. Normally in Indian reservoirs, with the objective of raising fingerlings in cage culture, water quality is not conducive to good fish health and on very rare occasions with a dense algal bloom, some parameters cause stress for the fish. Therefore, cages should be cleaned with soft coir brush fortnightly to remove biofouling organisms like algae, sponges and debris. Routine checking for loose twine, torn meshes from predators, anchors and sinkers is also necessary.

7.10. Fish Stock Monitoring

Prophylactic measures should be followed at least fortnightly and when necessary by lifting out fry inside the cages, soaking in salt solution (2-3 per cent), followed by permanganate solution (4-5 per cent) for two minutes to eradicate ecotoparasites and make the fry and aquatic environment more healthy. Also, a potassium permanganate solution (10-20 per cent) may be spread on water surfaces inside cages. At times a lime solution may be spread inside cages to clear water. A bath with malachite green at 0.001 per cent was used every 15 days to avoid any fungal attack in stocked crops. Fish health can be checked easily by noticing the feeding habits of fry when feed is applied. Regular sampling for growth assessment facilitates the feeding policy and harvesting schedule.

Economics of Production of Fry to Fingerlings in Cages

The capital investment for a battery of 8 cages, covering an area of 120 m^2 with a working volume of 320 m^3, required US$ 65.87/crop and a recurring cost of US$ 292.40/crop. So, production cost per crop including depreciation, interest, additional expenditure, devaluation of money was US$ 448.20. The market price of 70,000 fingerlings is US$ 1200, including transport, for the first two years and US$ 1496 for the next three years. Thus, the net benefit per crop of three months culture period is US$ 986.27. The benefit-to cost ratio is 2.20 and production cost per fingerling was estimated at US$ 0.0064. If the culture period of fingerling is reduced to two months, the production cost would be reduced further to US$ 0.0059/fingerling.

Economics of Fish Seed Raising in Cages (per crop), in Indian Rupees

Succeess and profitability of cage culture depends primarily on:

1. Selection of suitable body of water
2. Obtaining working facilities at reasonable cost
3. Stocking with quality fingerlings
4. Purchasing inputs economically
5. Achieving good fed conversion and
6. Maintaing high water quality

Initial Capital Costs

Capital requirements are difficult to estimate since the situation of each farmer is different.Equipment and facilitiues are also cpital items to be charged against the fish farming operations.Cages,whether built or purchased ready-made,should have frames with a life of atleast three years.The producer should have a small boat or coracle for checking,feeding and harvesting fish.Nets,feed buskets,tubs,a small stock of chemicals and instruments to monitor water quality are also necessary.

Production Costs

The two major components of costs in cage culture are the fingerlings and feed.Other costs include labour,interest on operating funds and repairs.Cost benefit analysis of the culture of catla (*Catla catla*) in 10 m^3 floating cages for six months grow out period has been worked out.

Investment Cost

Cage Construction

Net material (3 kg)	:	Rs.750
Conduit pipe (14 nos,each of 3.5 m long)	:	Rs.600
Welding and fabrication	:	Rs.250
HDPP cans(4 nos)	:	Rs.200
Twine,rope,feed tray,etc	:	Rs.600
Total	:	***Rs.2400***
Annual depreciation on cage cost (Useful life of 4 years)	:	Rs.600
Fingerlings (50 g each @ Re.1 per fingerling)	:	Rs.400
Feed (@ Rs.8 per kg for 840 kg)	:	Rs.6700
Total	:	***Rs.7700***

Returns

Fish production 280 kg (sale price @ Rs.35 per kg)	:	Rs.9800
Net profit	:	Rs.2100
Simple rate of return (Without interest on investment)	:	27 per cent

Depreciation

The life span of HDPE plastic nets was at least one year and with minor repair, it may be used for two years, depending on the level of management after harvesting. Net cages used for production of fingerlings for two years might be sold at 7-8 per cent of their initial procurement price. Hapa nets of 'Garware' make (mesh size 10-20 mm, multifilament, knotless) can be used for more than 5-6 years.

The lifespan of metal drums used as floats was at least 5-7 years and of this period, 10-12 crops of table-fish and 5-6 crops of fingerlings can be harvested. After seven years of use, drums can be sold at 40-50 per cent of their procurement price. Only 27 per cent of the surface area of each drum remained under water when used as floats and immediately after the harvest the inundated portion was painted after rotating the drum to expose the previously inundated surface. After one year, the whole drum must be repainted properly to extend its lifespan.

Bamboo used for framing normally lasts for two years, requiring a change of 10 per cent every year. The bamboo used in underwater frames lasts for more than three years with proper care. After use in cage frames, bamboo might be sold at a price of 5 per cent of the initial procurement. If a GI frame is used and maintained properly, a lifespan of more than 15 years can be expected. If a HDPE frame is used, the lifespan can exceed 10 years, provided proper maintenance, such as rotating the exposed portion during the next crop, putting the top portion under water. The thick nylon ropes used to anchor the cage battery can serve for more than 5 years, especially when they remain under water. The silk ropes for tying net cages to the bamboo frame lasts for at least two years.

Constraints of Cage Culture

Unless managed carefully and correctly, cage culture can have certain drawbacks. Cages occupy space on the surface of water bodies and, if poorly positioned, may disrupt navigation or diminish the scenic value of the reservoir. Poorly placed cages may alter current flows and worsen sedimentation. Inappropriately intensive or poorly managed cage culture may pollute the environment with unconsumed feed and fish faecal waste, causing eutrophication. During the summer months, cages may be damaged by strong winds or flooding, but this risk can be avoided by properly anchoring batteries of cages in protected inlets away from strong currents. Theft is rarely a problem in culturing fish to fingerling size. As particular problems exist where intensive cage aquaculture is practised for producing marketable fish or prawns, the technology is used in Indian reservoirs only to rear fingerlings, with limited use of supplemental feed. As most of the reservoirs are in transition from an oligotrophic to a mesotrophic state, with very few in danger of eutrophication, the controlled discharge of waste products from cage culture can be immensely helpful in maintaining water nutrient levels.

Conclusion and Prospects

Despite the unsuitability of wetlands for cage culture, recent studies conducted by CIFRI indicate that raising fingerlings from cages is profitable and, if practiced in Indian reservoirs, could solve the problem of providing fingerlings for stocking those reservoirs. Furthermore, with prolonged, sustained and continued effort, assimilating the lessons learned, CIFRI achieved success in producing fingerlings in floating cages installed in Pahuj Reservoir (U.P.) and Dahod Reservoir (M.P.) during 2007-2009 (Bene *et al.,* 2009). In 2011, CIFRI ventured into raising table-fish of economically important species through cage culture by installing cage facilities with durable galvanized iron frames in Maithon Reservoir (Jharkhand) in 2011.

Many modifications and combinations were in practice to increase the cage viability, such as:

1. Use of different netting material for long-lasting durability without impacting fish growth,
2. Reduction of the cost of cage installation by replacement of simple pipe with galvanized iron pipe,
3. Replacement of plastic pipe with painted iron pipe to withstand strong wind action, making it long lasting and for antifouling, and
4. Provision of flexibility to better maneuver cages.

One of the prime objectives of cage culture discussed here is to recommend the rearing of fingerlings >100 mm in length, especially carps, for stocking reservoirs. Nevertheless, a general thinking has arisen among planners to depend more on open waters and culture-based capture fisheries to achieve high fish production through ecosystem management. Against this backdrop, the wide applicability of modified technologies, such as cage culture to enhance fish production, should be adopted. However, more comprehensive investigation is needed to characterize its impact on ecosystems and the sustainable accumulation of capacity for auto-stocking of indigenous fishes in inland open-water systems in general and reservoirs in particular.

Concept of Integrated Multi-trophic Aquaculture (IMTA)

Integrated aquaculture systems as detailed in Neori *et al.* (2004) Barrington *et al.* (2009) and Angel and Freeman (2009) and case studies in India are briefly given below. In many monoculture farming systems the fed-aquaculture species and the organic/inorganic extractive aquaculture species (bivalves, herbivorous fishes and aquatic plants) are independently farmed in different geographical locations, resulting in pronounced shift in the environmental processes. Integrated multi-trophic aquaculture (IMTA) involves cultivating fed species with extractive species that utilize the inorganic and organic wastes from aquaculture for their growth. According to Barrington (2009), IMTA is the practice which combines, in the appropriate proportions, the cultivation of fed aquaculture species (*e.g.* finfish/shrimp) with organic extractive aquaculture species (*e.g.* shellfish/herbivorous fish) and inorganic extractive aquaculture species (*e.g.* seaweed) to create balanced systems for environmental sustainability (biomitigation) economic stability (product diversification and risk reduction) and social acceptability (better management practices). This farming method is different from finfish "polyculture", where the fishes share the same biological and chemical processes which could potentially lead to shift in ecosystem. Multi-trophic refers to the combination of species from different trophic levels in the same system. The multi-trophic sub-systems are integrated in IMTA that refers to the more intensive cultivation of the different species in proximity of each other, linked by nutrient and energy transfer through water.

7.11. History of Pen Culture

Compared to cage culture,pen culture has a more recent history.It possibly started in the Inland sea area in Japan in the early 1920s and later spread to China in the early 1950s for rearing carps in freshwater lakes.Pen culture was taken up on a commercial scale in Laguna de Bay and the San Pablo lakes in the Phillipines from 1968 for rearing milk fish, chanos chanos.

Presently, commercial pen culture of fish is in vogue in the Phillipines,Indonesia and China.The main species cultured in pens in these countries are milk fish,grass,big head and silver carps,and tilapia,In India in situ carp seed rearing was attempted in Bhavani sagar and Tungabhadra reservoirs in pens since 1978 for rearing fingerlings.

Pens are still constructed much the same way as they always were,except that nylon or polythene-mesh nets have replaced the traditional split bamboo fences. The nets are attached to fixed posts set every few meters and the bottom of the net is buries ion the substrate with long wooden pegs.

7.12. Pen Culture

Fish culture in pens has a short history compared to cage culture. It has been experimented in Japan during early 1920s. Pens are used for various purposes such as holding the fish temporarily for a short period before shifting to other places, raising fish for table purpose, short term rearing of fin-fish and shell-fishh, *etc.* in countries such as Philippines, Indonesia, Thailand, Malaysia, China and U.S.A. In India, pen culture has been experimented in an oxbow lake in Bihar (Banerjee and Pandey, 1978), Beels in Assam (Yadava *et al.,* 1983) and swampy tank at Bhavanisagar in Tamil Nadu (Abraham, 1980) for raising carp seed. Similarly, carp spawn produced in excess than the capacity of fish farm at Tungabhadra dam is cultured in pens erected in protected bay area of the reservoir since 1982 (Swaminathan and Singit, 1982). Pens assume importance as the material required for fabrication is cheaper, readily available in the market and even unskilled personnel can erect them without involving engineering skill. The pen structure can also be relocated at other suitable sites. The material may last for 3 to 4 years for raising several crops. Thus, pen culture can serve as one of the cheaper alternatives to costly land-based rearing space. Fish pen designing and construction is easy and simple when compared to that of cages. They are of various shapes (circular, semi-circular, square and rectangular). Rectangular shape will be convenient for harvesting, using drag net. The size of the pen can be from 10 m2 to 5 ha area. However, smaller units can be easily managed. The height of the pen wall depends on the water level during the culture period. It will be convenient to have a depth ranging from 1 to 3 m. Pen is defined as "a fixed enclosure in which the bottom is the bed of the water body". It involve rearing of fish within fixed enclosures supported by frameworks of bamboo, wood or metal, set in sheltered, shallow portions of lakes, bays, rivers, and estuaries.

7.13. Seed Production in Pens

Production of seed in pens is easier than in cages, as the pens are installed in the marginal area of the reservoir where of freshwater exist. The mesh size can be as large as possible for better water circulation, ensuring that brood fish do not escape. A water depth of 60 to 100 cm can be maintained in the pens meant for spawning. Brood fish consisting of gravid males and females are introduced into the pen for spawning and the eggs are collected for hatching elsewhere. Spawning of tilapia in pens is a regular practice in lakes in Philippines.

Rearing of Spawn

Split bamboo strips are woven into a mat with nylon twines and it is provided with fine meshed velon screen material and the pen is erected as described above.

Spawn of desired fish is stocked in the pens and reared for raising fry and fingerlings. Unlike cage culture, all the management measures, which are followed in land-based nurseries, must be followed in pen culture also. The pens must be fertilized with organic or inorganic manure. Unwanted insects and enemies of fish spawn and fry must be eliminated.

Supplementary feeding with balanced diet must be provided in feeding trays.

Assessment of growth, survival and health conditions must be monitored regularly. Harvesting is done using seine net.

Rearing of Fry to Fingerlings

The mesh size of the screen material for rearing fry to fingerlings may be slightly bigger (2 to 4 mm). Raising of food fish in Pens: Poly-culture can be followed in pens to exploit all feeding niches therein. Species selection, composition and stocking rates will depend largely on the natural food supply, supplemental feeding, water depth and duration of water availability in the site. 5-10 fingerlings per sq. m may be followed for carps with supplemental feeding. In Bihar a pen of 120 m^2 was stocked with catla, rohu and mrigal in the ratio of 7: 12: 16 per m2. The fish were fed with a mixture of rice bran and mustard cake (2: 1 ratio) at 5- 10 per cent body weight. They attained an average weight of 1100, 800 and 750 g from an initial weight of 166, 75 and 120 g respectively and yielded a computed production of 25t/ha/6 months.

Setting up of Fish Pen

Suitable sizes of fish pens for single-family operations are 0.3-0.5 ha.Ready-made netting or bamboo screen is fixed to the bamboo posts driven 0.5-1 m deep into the bottom soil to form a circular or rectangular enclosure.Mesh size of net or bamboo screen should be small.Round fish pens use less materials per unit area. The top of fish pen should be 0.5-1 m above the expected maximum water level. The bottom should be buried 20-30 cm into the soil using a weight or sinker. The pen enclosure should have a gate to allow the boat to go in and out the pen for production operations.

Types of Pen

1. Completely isolated enclosures surrounded by a net structure in the middle of a bay with no foreshore;
2. Shore enclosure with a portion of the foreshore extending into deep water surrounded by a net structure, and
3. Bay or lock enclosure with an embankment provided with sluices or net structure only at the entrance

Classification of Pen

Based on design and construction:

1. Rigid pens
   Embanked pens
   Net enclosures
2. Flexible pens (netting)
3. Outer barrier nets. Based on feeding:
   Extensive,
   semi-intensive,
   Intensive

Economics of Pen Culture in Indian Reservoir

A. Capital cost

Net material (3 kg)	:	Rs.750
Cost of casuarina poles	:	Rs. 17600.00
Cost of bamboo poles	:	Rs. 6000.00
Cost of HDPE woven fabrics	:	Rs. 14000.00
Stitching charges	:	Rs. 1000.00
Total	***:***	***Rs. 38600.00***

B. Capital cost per crop

(25 per cent of A - assuming that the material will go for 4 crops)	:	Rs. 9850.00

C. Recurring cost

Earth work (trench making)	:	Rs. 2000.00
Synthetic and coir ropes, nylon twine, etc.	:	Rs. 500.00
Installation charges	:	Rs. 1500.00
Cost of seed	:	Rs. 8400.00
Cost of feed	:	Rs. 3100.00
Watch and ward	:	Rs. 3000.00
Netting and harvesting charges	:	Rs. 1000.00
Total	***:***	***Rs. 20500.00***
D. Total cost (B+C)	***:***	***Rs. 30150.00***
E. Estimated sale receipts	***:***	***Rs. 38059.00***
F. Profit	***:***	***Rs. 7909.00***
G. Return on investment		***26.2 per cent***

7.14. Economics of Pangasius Cage Culture

Rearing fingerlings from fry using cage culture in the reservoirs to be stocked is more cost-effective than using either pens or nurseries. One battery of eight cages is sufficient to produce stocking materials for a water body of 200 ha. Three crops of fingerlings can easily be harvested per year.

The useful life of HDPE plasto nets is at least 1 year and, with minor repair, may extend to 2 years, depending on their management after each harvest. Nets should be cleaned immediately after the harvest, dried in the sun, and either stored properly or immediately reused. After a year of use, Nelton may be sold as scrap for 7-8 per cent of the initial procurement price.

As steel drums used as floats have only 27 per cent of their surface area under water, they can easily be rotated immediately after the harvest to allow repainting the previously submerged portion. To maximize the useful life of the drums, the whole drum should be repainted annually. Steel drums used as floats can be used for at least 5 years, during which time 15 fingerling crops can be harvested. After 5 years of use, they can be sold for 40 per cent of their procurement price.

Bamboo poles used for frames normally last for 2 years. With 10 per cent of poles needing to be changed every year. The poles used in the submerged frame will last for more than 3 years with proper care and management. Used bamboo poles may be sold for 5 per cent of their initial price.

The thicker nylon ropes used for anchors serve for more than years, especially the portions that remain under water. The silk ropes for typing Nelton cages with bamboo frame last for at least 2 years.

Cost of Production of Each Crop

The capital investment for a battery of eight cages covering 120 m^2 with a working volume of 320 m^3 is Rs. 4,117 per crop. Recurring expenses come to Rs. 18275 per crop. This yields a production cost per crop – allowing for depreciation, interest, additional expenditures, and inflation of Rs. 28,013. The market price of 70,000 fingerlings is Rs. 75000 including transport for the first 2 years and Rs. 93,500 for the remaining 3 years. The income per crop, cultured over 3 month is Rs. 61,642. This yields a highly favourable cost benefit ration of 2:20, with the production cost per fingerlings estimated at Rs. 0.50. If the fingerlings culture period is reduced to 2 months, the production cost drops to Rs. 0.37, enhancing the cost: benefit ratio to 2:25.

The detailed economics of cage culture for a grow-out period of 3 months per crop at the price prevailing is given in Table 31.

Table 31

	Size/Weight	*No./kg*	*Unit Cost (Rs.)*	*Life (Years)*	*Depreciation (Years)*	*Total Cost (Rs.)*	*Cost/Crop (Rs.)*
1. Capital cost							
A. Cage culture							
Steel Drums	200 litre	24	550	5	70	1,680	560
Bamboo Poles	7.5 metre	56	110	3	30	1,680	560
Transport		1 Truck	800	5	160	160	53
Measuring tape	Large	1	100	10	10	10	3
Nuts and Bolts	18 cm	120.9 kg	60	5	11	99	33
GI Wire	2 mm	20 kg	70	1	70	1,400	466
GI Wire	1.5 mm	25kg	70	1	70	1,750	583
Pliers	Medium	2	70	2	35	70	23
Bucket, mug and strainer	Large	2 each	70+10+10	5	14+2+2	36	12
Cutter	Medium	2	50	10	5	10	4
Paint	Primer	3 kg	50	2	25	75	25
Paint	Acrylic	4 kg	120	2	60	240	80
Labour	Painting	8	100	1	100	800	266
Labour	Framing	8	100	1	100	800	266
Frame Subtotal							**2,934**

	Size/Weight	No./Kg	Unit Cost (Rs)	Life (Years)	Depreciation (Years)	Total Cost (Rs)	Cost/Crop (Rs)
B. Netting							
Net cage	45 m^3, 1.5mm	8	800	2	350	2,800	933
Silk rope	1.27 cm diam.	5kg	200	5	40	200	66
Nylon rope	1.27 cm diam.	10kg	70	3	23	230	77
Nylon rope	2.5 cm diam.	20kg	80	5	16	320	107
Netting Subtotal							**1,181**
Total Capital Cost (A+B)							**4,117**
2. Recurring Costs							
Labour	Net tying	5	100	1/3	300	1,500	**500**
Fry	12-18	10,000	0.12	1/3	3.6	36,000	**12,000**
Feed	Dust, flaky	270 kg	12	1/3	36	9,720	**3,240**
Agrimin nutri. Supplement	Dust, flaky	3kg	45	1/3	135	405	**135**
Labour	Daily feeding	1	1,000/month	1/3	9,000	9,000	**1,000**
Labour	Cleaning, *etc.*,	10	100	1/3	300	3,000	**1,000**
Labour	Harvest	4	100	1/3	300	1,200	**400**
Total recurring cost per crop							**18,275**

Production cost per crop, including interest, additional expenditure and inflation is Rs. 28,013.Income: market price of 70,000 fingerlings @ Rs. 1.00/fingerlings is Rs. 75,000 including transport, for the first 2 years and @Rs. 1.25/fingerling is Rs. 93,500, including transport for the remaining 3 years. Thus, income is Rs. 89,655.Expenditure: Rs. 28,013; Benefit: Rs. 61,642 and benefit: Cost = 2.20; production cost of one fingerling: Rs. 0.40.

Table 32: The Material Requirements and Economics of Fish Seed Raising in Cages in Indian Reservoirs

Item/Particular	*Dimension*	*Unit*	*Quantity*	*Rate (Rs.)*	*Amount (Rs.)*	*Life (Years)*	*Depreciation (Rs.)*	*Cost per Crop @ 2 Crops/ Year (Rs.)*	*Cost per Crop @ 3 Crops/ Year (Rs.)*
Nonrecurring costs									
Capital costs									
Floats (PVC drums)	200 litre	Number	24	550	13,200	10	1,320	660.00	440
Bamboo	7.5 meter	Number	56	110	6,160	5	1232	616.00	410.67
Nuts and bolts	2mm	No./kg	120/g	60	540	5	108	54.00	36.00
Iron Wire	18	kg	20	70	1,400	1	1,400	700.00	466.67
Iron Wire	1.5mm	kg	25	70	1,750	1	1,750	875.00	583.33
Saw	Medium	Number	2	50	100	10	10	5.00	3.33
Measuring tape	Large	Number	1	100	100	10	10	5.00	3.33
Cutting pliers	Medium	Number	2	70	140	2	70	35.00	23.33
Net Cage	$45m^3$ 1.5m	kg	8	800	6,400	2	3,200	1,600.00	1,066.67
Silk rope (diameter)	1.27 cm	kg	5	200	1,000	5	200	100.00	66.67
Nylon rope (diameter)	1.27cm	kg	10	70	700	3	233	116.67	77.67
Nylon rope (diameter)	2.25	kg	20	80	1,600	5	320	160.00	106.67
Bucket, mug and strainer		Number	2 each	45	90	5	18	9.00	6.00
Paint	Primer	kg	3	50	150	2	75	37.50	25.00
Paint	Acrylic	kg	4	120	480	2	240	120.00	80.00
Fixed Capital Cost							**10,186.33**	**5,093.17**	**3,395.44**
Labour for frame and repair	Day	Person day	8	100			800	200.00	133.33
Labour for netting and repair	Day	Person day	5	100			500	125.00	83.33
Labour for paint and repair	Day	Person day	8	100			800	200.00	133.33
Interest on fixed capital		Rs					982.91	449.45	299.63
Total fixed cost		**Rs.**						**6,067.62**	**4,045.07**

Item/Particular	Dimension	Unit	Quantity	Rate (Rs.)	Amount (Rs.)	Life (Years)	Depreciation (Rs.)	Cost per Crop @ 2 Crops/ Year (Rs.)	Cost per Crop @ 3 Crops/ Year (Rs.)
Recurring Costs									
Seed	12-18mm	Number	100,000	0.12				12,000	12,000
Feed	Dusk, Flaky	kg	270	12				3,240	3,240
Agrimin	Dusk, Flaky	kg	3	45				135	135
Transportation	Number	Truck	1	800				800	800
Feeding	Month	Person-day	3	1,000				3,000	3,000
Cleaning	Day	Person-day	10	100				1,000	1,000
Harvesting	Day	Person-day	4	100				400	400
Miscellanious		Rs.						500	500
Total Expenditure		Rs.						21,075	21,075
Interest on variable cost		Rs.						1,264.50	1,264.50
Total Variable cost		Rs.						22,339.50	22,339.50
Total cost		Rs.						28,407.12	26,384.57
Fingerling production	70-100mm	Number	70,000	1	70,000				
Cost of Production	Per fingerling	Rs.						0.41	0.38
Value of fingerlings		Rs.		1	70,000			70,000	70,000
Benefit: Cost ration								2.46	2.65

Table 33: Economics of Fish Seed reading in cages In Indian reservoirs.

Item	*Per Crop @ 2 Crops/Year (Rs)*	*Per Crop @ 3 Crops/year (Rs)*
Total fixed cost	6,067	4,045.07
(Per cent of total cost)	(21.36)	(15.33)
Total variable cost	22,339.50	22,339.50
(Per cent of total cost)	(78.64)	(84.67)
Total cost	28,407.12	26,384.57
Number of fingerlings produced	70,000	70,000
Cost of production/fingerling	0.41	0.38
Value of fingerlings @ Re. 1/fingerlings	70,000	70,000
Benefit: cost ratio	2.46	2.65

Chapter 8

Present Status and Future Prospects of Aquaculture in Reservoir

8.1. Fish Production Trends and Potential

The fish production statistics from inland sources are inaccurate all over the world due to many reasons and reservoirs are no exception to this. Nevertheless, it has been established that the fish yield from Indian reservoirs is rather poor varying from 0.05kg/ha in Bihar to 35.5 kg/ha in Himachal Pradesh with national average of 20 kg/ha. The average national yield from small reservoirs in India is nearly 50-kg/ha, which is also low (3.9 kg/ha in Bihar to 188 kg/ha in Andhra Pradesh), which is well below the rates achieved in many other countries such as >800 kg in China, 300 Kg in Sri Lanka and 100 kg/ha in Cuba.

8.2. Fish Production Potential

Prioritizing culture-based fisheries of reservoirs holds the key for increasing inland fish production in India. Based on the average fish yields of 422 reservoirs, it has been estimated that small, medium and large reservoirs yield fish at the rate of 49.05 kg/ha, 12.30 kg/ha and 11.43 kg/ha respectively. Applying this at national scale, fish production from 1,485,557 ha of small, 507,298 ha of medium and 1,160,511ha of large reservoirs, could be 74,129 tonnes, 6,488 tonnes and 13,033 tonnes from small, medium and large reservoirs respectively. But, with moderate increased yield rate of 500,250 and 100 kg/ha for small, medium and large reservoirs, respectively, the production is expected to be 743,000, 127,000 and 116,000tonnes for small, medium and large reservoirs respectively. Cumulatively this would boost up the present production of all reservoirs from the current 64,00tonnes to nearly 1 million tones.

8.3 Impact of Reservoir Formation on the Native Ichthyofauna

Formation of reservoirs have affected especially the following Indigenous fish stocks:

1. The mahseers, snow trouts and *Labeo dero* and *L. dyocheilus* of the Himalayan streams.
2. The anadromous hilsa, the catadromous eels,and freshwater prawns of all major river systems.
3. *P. sarana, T. tor, Tor mahanadicus, T. mosal, L. fimbrtatus, L. calbasu* and *Rhinomugil corsula* of the Mahanadi river.
4. *P. dobsoni, P. dubius, P. carnaticus, C. drrhosa* and *Labeo kontius* of the Cauvery basin.
5. *P. kolus, P. dubius, P. sarana, P. porcellus, L.fimbrtatus, L. calbasu, L. pangusia* and *Tor kudree of the Krishna river system*, and
6. The mahseers, eels and *Osteobrama belangiri* of the north-east (Figure 16).

The pristine streams of the river Sutlej harboured at least 51 species of fishes including the (exotic) trout, *Salmo trutafario*, the snow trouts, *Schizothorax* spp. and several species of hillstream fishes (Anon. 1989b). Most of them were unique due to the sub-temperate climate and the zoogeographic affiliations to the Himalayan region. The upper reaches of the Sutlej and its tributaries were particularly rich in *T. putltora, L. dero, L dyocheilus* and *Schizothorax* spp. A decline in the number of species and their populations has been reported on account of changed ecological conditions especially the silt deposition at the bottom and the stratification of water body. Apart from minnows, many native species, such as, *Schizothorax plagiostomus, T. putitora, L. dero, L. dyocheilus* are on the decline. Proliferation of the exotic carps, *H. molitrix* and *C. carpio* in recent year has further to the local species.

Before the creation of Hirakud reservoir, the parent river Mahanadi had a rich fish fauna of 103 species, comprising both plain and sub-montane forms with sizeable representation of carps and catfishes. The common species of the river were *P. sarana, T. tor, T. mahanadicus T. mosal, L. fimbriatus* and the Indo-Gangetic major carps. The endangered *T. mosal* and *T. mahanadicus* were protected in the temple tanks that were submerged during reservoir formation. Presently, the number of species has declined to 40, of which many may disappear. The worst affected are *Tor mosal, Rhinomugil corsula* and the freshwater prawn *Macrobrachium malcolmsonii.*

Fishes affected in the two reservoirs in the Krishna river system *viz.,* Tungabhadra and Nagarjunasagar are *P. sarana*, and *Labeo* spp. Soon after the impoundment, the Tungabhadra reservoir harboured a good population of indigenous *P. kolus* that contributed up to a third of the total fish landings. *P. dubius, P. sarana, P. porcellus, P. potail* and *Labeo pangusia* were also present in large numbers. Most of these native species have either disappeared or declined drastically due to the absence of fluviatile environment and the changed trophic

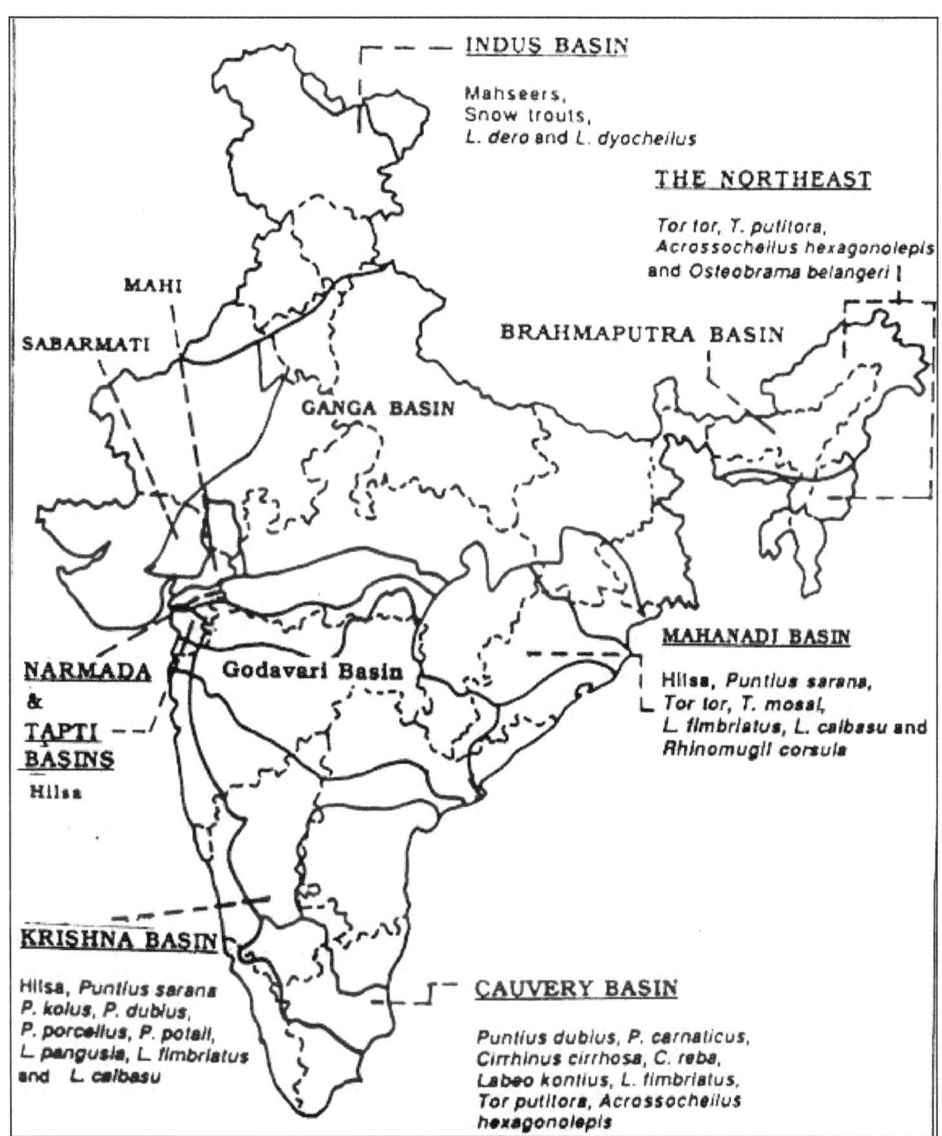

Figure 20: Indigenous Fish Species Affected by Reservoir Formation in India.

structure. The vacant niche has been filled by minnow-predator combination, due to the management lapse of not inducting the fast-growing species into the system.

Nagarjunasagar reservoir on the mainstream Krishna harboured rich populations of *Labeo fimbriatus, Labeo calbasu* and *T. khudree* in the earlier years of impoundment. On account of recurring breeding failure and habitat loss, these species have declined over the years, giving way to minnows which have shared the common niche with them. In the absence of any commercial fish to utilize the rich planktonic resources of the reservoir, they are mainly channelled through the detritus-molluscs chain to favour *Pangasius pangasius* and through grazing-predator chain to help the *Silonia childreni* populations. Thus, finally, these two catfishes have established a firm hold on the fish fauna.

Cauvery river system is the original abode of a number of fish species including *P. dubius, P. carnaticus, C. cirrosa, C. reba, L. kontius, L. fimbriatus, Tor putitora* and *Acrossocheilus hexagonolepis* which have been affected in various degrees by the impoundments, chiefly, Krishnarajasagar, Mettur, Bhavanisagar and Amaravathy reservoirs. During the last 25 years, the indigenous economic carps in Krishnarajasagar *viz., Labeo spp., P. dubius* and *P. carnaticus* have suffered setback due to the changed ecological conditions, especially the components of fish food biotic communities. These species, contributing more than 60 per cent of the total catch during the 1950s, have given way to the transplanted exotic common carp, *C. carpio*, which found a favourable environment in the reservoir *vis-a-vis* feeding and breeding. At present, most of the energy transfer is channelled through the detritus/benthos chain, giving considerable edge to the common carp which is a prolific breeder and competitor to *Cirrhinus* spp. for food.

One of the earliest casualties of hydraulic structures is the Indian shad *Tenualosa ilisha* (the hilsa) which was affected as early as mid 19th century when the Upper and Lower Anicuts were constructed on Cauvery. These barrages had severly restricted the migration of hilsa by obstructing their pathways and construction of Mettur dam (Stanley reservoir) in 1935 completely stopped the hilsa run in Cauvery. Several fishes were affected by the Mettur dam. *Puntius* spp. which used to form 28 per cent of the landings in 1943–44 faded out in the mid 1970s. Although the indigenous *Cirrhinus cirrhosa* took some initial advantage, it also could not survive. Low water levels during July for three consecutive years have most probably caused its decline. Similarly, *Labeo kontius* which was next only to *C. cirrhosa* in Cauvery also disappeared from the reservoir. The Gangetic carps transplanted into the reservoir also did not find roots there. *C. mrigala* was stocked in 1950–51 and appeared in catch during 1957–58, contributing up to 13.9 per cent in 1966–67, only to fade into insignificance later. The some fate met *Labeo rohita*. Recruitment failure, water level changes, and predator pressure are the main reasons for the failure of Indian major carps in Stanley reservoir. In 1993, the total catch of 115 t in the reservoir comprised *L. rohita (19 per cent), Wallago attu* (15 per cent), other catfishes (14 per cent), *Puntius spp* (14 per cent) and *C. catla* (10 per cent).

Bhavanisagar is the only reservoir in the Cauvery basin, where the indigenous species like *Puntius* spp., *Tor putitora, T. tor, A. hexagonolepis, P. dubius, P. carnaticus, L. kontius* and *C. cirrhosa* still hold together well. Their survival is mainly due to the uninterrupted breeding activities at Moolathurai and Nellithurai, especially when water is released from the upstream Pilloor reservoir. *P. dubius* ascends the river Moyar during the north-east monsoon and lays eggs in batches of 1,000 to 2,000 on the gravel beds. Similar breeding success has been confirmed in case of *Cirrhinus reba, Labeo fimbriatus, Labeo calbasu, L. kontius* and *Puntius carnaticus.*

Amaravathy and Sathanur with their prime fishes of tilapia and catla respectively are the examples of introduced fishes finding a favourable environment and propagating themselves into a dominant position.

8.4. Positive Impact of Reservoirs on Fish Fauna

Many species of fish not only manage to adapt to the reservoir ecosystem but also find it congenial and flourish there, which is the main reason for the rise in the biomass of reservoir in the early stage of impoundment. However, most of the fishes that manage to multiply in the reservoir system are not very high in priority from the commercial and ecological point of view. Stocks of the small clupeid, *Salmostoma phulo phulo* and *o. vigorsii*, which support a flourishing dry fish trade in Nagarjunasagar and Tungabhadra reservoirs, multiply in a much higher scale than they do in the riverine ecosystem. The catfish, *P. pangasius* which was believed to be a catadromous migrant, has not only adapted itself to become a resident population in Nagarjunasagar, it has also become a very important component of the population. Ramakrishniah (1994) described many instances where reservoirs acted as sanctuaries by citing examples of *Barilius bola* in Tilaiya (Damodar), *Mystus krishnensis, Osteobrama vigorsii*, and *Pseudeutropius taackree* in Nagarjunasagar (Krishna), *T. sandkhol* in Nizamsagar (Godavari), *Tor khudree* and *T. mussullah* in Shivajisagar (Krishna), *A. seenghala* and *T. putitora* in Pong (Beas) and Vallabhsagar (Tapti).

Modelling is a scientific approach to bring ecosystem approaches for fisheries management. Ecopath is a trophical modelling approach which is used to model a wide variety of aquatic ecosystem. Ecopath is a component of Ecopath with Ecosim software, which is a free modelling software suite. The aggregation of the species into functional groups is essential to perform the network analysis and by providing the basic input parameters along with it, the model helps to estimate the ecosystem indicators which help to better explain the ecosystem. The Ecopath model, a mass balance model is really helpful in understanding interactions among the different ecological components and which can be well utilised in developing strategies for the management of aquatic ecosystem.

8.5. Integrated Approach in Indian Reservoirs

Very small reservoirs are amenable for integrated aquaculture since the culture based fishery can be effectively combined with piggery, duckery and poultry rearing.

Many of the waste products from these animal husbandry practices act as fish food or fertilizer leading to higher fish yield. Such systems have special relevance to the small reservoirs of the north east.However, this approach has limitations from aesthetic and hygienic point of view, especially when the reservoir is a source of drinking water supply. The exposed areas of the reservoir can be auctioned for agricultural farming of leguminous crops, which would also add to the productivity of the soil.Such increase in fish production and earnings can make a significant contribution to the nutritional requirements of the rural community.

Aquaponics

Combining aquaculture (farming aquatic species) and hydroponics (soil-less plant culture) with an input of fish feed, nitrogen-rich fish wastes are converted by bacteria into nutrients for growing plants that in turn biologically filter the water.

This method farming fish and crops is a good thing on several different levels. First of all it removes fertilizer and chemicals from the agricultural process.The fish wastes act as a natural fertilizer for the crops, instead. Second of all, it saves water because the water is recycled within the tanks rather than sprayed across can be setup anywhere,so it reduces the communities fish and crops a field of crops with abdomen.Thirdly,an aquaponics environment need for local communities.

Aquaponics is a method of cultivating both crops and fish in a controlled environment. The fish are kept in tanks and the plants are grown hydroponically (without soil). They sit in beds, but their roots hang down into a tub of water. When fish live in tanks, their wastes builds up in the water, and it eventually becomes poisonous of them. But what is toxic for fishis nourishing for plants, they love nothing more than to suck down some fish waste-laden water from the fish tanks is funnelled to the tubs where the plants dangle their roots.When the plants absorb the nutrients they need from that water, they basically cleanse it of toxins for the fish. Then that same cleansed water can be funnelled back into the fish tanks.

Advantages

1. Significant reduction in usage of water
2. Growth of plants is significantly faster
3. Vegetables are bigger and healthier
4. No need to use artificial fertilizer
5. No need to dispose of fish water or provide an artificial filtration
6. Reduction in land is required to grow the same crops
7. Easier to setup for year round use.

Glossary

Alkalinity – A measure of the capacity of water to neutralize a strong acid. In natural waters this capacity is attributable to basics ions such as bicarbonate, carbonate and hydroxyl ions as well as other ions often present in small concentrations such as silicates and borates.

Aquaculture – The farming of aquatic organism including fish, molluscs crustaceans and aquatic plants with some sort of intervention in the rearing process to enhance production, such as regular stocking, feeding, protection from predators, *etc.* Farming also implies individual or corporate ownership of the stock is being cultivated.

Aquatic organisms – A complex of living organism and their non-living environment, which are inseparably interrelated and interact upon each other.

Bathymetry – The measurements defining the size, depth and shape of a lake.

Benthos – Animal attached to, crawling or burrowing into the bottom substrata in a water body with no or limited mobility.

Biological diversity – The variety and variability among living organisms from all sources including, *inter alia*, terrestrial, marine and other aquatic ecosystems and the ecological complexes of which they are part; this includes diversity within species, between species and of ecosystems.

Biological indicator – Organism, species or community whose characteristics show the presence of specific environmental conditions. Other terms used are indicator plant and indicator species.

Biological interaction – An interaction between species or stock elements resulting from direct predation or competition for food (predator-prey relationships) or space (or both). Because of their importance, the impact of fishing on associated or dependent species needs to be assessed.

Biomass – Also referred to as the standing stock. The total weight of a group (or stock) of living organisms (*e.g.* fish. Plankton) or some defined fraction of it (*e.g.* spawners), in an area at particular time.

Biotic – Live and living organisms.

Bloom – When related to phytoplankton, a sudden and rapid increase in biomass of the plankton population. Seasonal blooms are essential for the aquatic system productivity. Sporadic plankton blooms can be toxic.

Brood stock – Specimen or species, either as eggs, juveniles, or adults, from which a first or subsequent generation may be produced in capacity, whether for growing as aquaculture or for release to the wild for stock enhancement.

Capture-based-aquaculture – Aquaculture system where seed is sourced from natural (wild) collections.

Carrying capacity – The maximum population of a species that a specific ecosystem can support on a sustainable basis.

Catchment area – The area drained by a river or body of water.

Co-management – A governance system in which representatives from the community and government participate in decision-making processes.

Common property resource – A resource held collectively and managed by a community or a particular group (two or more persons) within a community for the common benefit. Excludes individual rights.

Conservation – Actions, which are directed towards sustaining otherwise decreasing rates of use, towards sustained yield management, or towards increasing a sustained use.

Culture-based fisheries – Fisheries that depend solely or mostly on stocking (stock and recapture).

Draw-down – The lowering of a reservoir or lake by controlled withdrawal.

Detritus – Tiny particles of material found in sediment or suspended in water organic detritus is derived from the decomposition of organisms: inorganic detritus is derived from the erosion of rocks and other mineral materials.

Ecological impact – Effect of human activities and natural events on living organisms and their non-living environment.

Ecosystem – System of plants, animal and microorganisms together with the non-living components of their environment.

Ecosystem-based management – An approach that takes major ecosystem components and services-both structural and functional-into account in managing fisheries. It values habitat, embraces a multi-species perspective, and is committed to understanding ecosystem processes. Its goal is to rebuild and sustain populations, species, biological communities and ecosystem at high levels of productivity and biological diversity so as not to jeopardise a wide range of goods and services from ecosystems while providing food, revenues and recreation for humans.

Edaphic factors – Water and soil quality parameters that has a bearing on fish productivity.

Enhancement – Qualitative and quantitative improvement in productivity of water bodies through specific management options.

Enhancement through new culture system – Cage culture, pen culture, Fish Aggregating Devices (FADs) *etc.*

Environment – The combined external conditions affecting the life, development and survival of an organism or an ecosystem.

Environmental enhancement – Input of nutrient through fertilization to increase plankton of enhanced fish production in reservoirs.

Environmental impact – Direct effect of human activities and natural events on the components of the environment.

Environment impact assessment – A sequential set of activities deigned to identity and predict the impacts of a proposed action on the bio-geophysical environment and on man's health and well being, and interpret and communicate information about the impacts, including mitigations measures that are likely to eliminate the risks.

Epilimnion – Water column below the thermocline.

Eutrophication – Natural or man-induced process by which a body of water becomes enriched in dissolved mineral nutrients (particularly phosphorus and nitrogen) that stimulate the growth of aquatic plants and enhance organic production of the water body.

Exotic species – Specific not native to a particular area.

Fish Aggregating Devices (FADs) – Artificial substrata provided in natural water bodies fish to take shelter for their easy capture.

Fish production – Total quantity of fish produced from a water body.

Fish yield – Quantity of fish produced from a unit area and time.

Food chain – Organisation of plant and animal communities into groups, which are dependent on one another for food. These communities from a tropic chain that transforms energy from one level to the other.

Forage species – Species used as prey by a predator for its food.

Habitat – The environment in which the fish live, including everything that surrounds and affects its life: *e.g.*, water quality, bottom, vegetarian, associated species (including food supplies).

HDPP – High Density Polypropylene.

Hypolimnion – Water column above the thermocline.

Indigenous species – A species living in its natural of distribution.

Integrated production system – Combining fish production with crop and animal husbandry particles to enhance benefits.

Isotherms – Lower animals (lower than birds and mammals) that do not maintain their own body temperature. It is determined by the ambient temperature of the surroundings.

Limnology – Study of the ecology of freshwater ecosystems.

Lentic ecosystem – Ecosystem of a lake, pond or swamp.

Littoral vegetation – Plants, mostly terrestrial adapted to living in submerged conditions at the interface between land and water.

Lotic ecosystem – Ecosystem of a river, stream or spring.

Microvegetation/Macrophytes – Submerged, floating, rooted. Emergent or rooted and emergent aquatic plants.

Management enhancement – Introducing new management option like recreational fishery, *etc.*

Mortality – Depletion of stock from a population. Depletion due to natural causes is natural mortality and depletion due to fishing is fishing mortality.

Niche – The role and position of a species in an ecosystem.

Oligotrophic – Water bodies deficient in nutrients.

Open access – A condition of a fisheries in which anyone who wishes to fish may do so.

Periphyton – Organisms (mostly microscopic), both plants and animals attached to a submerged substrata like stones, pebbles, plant trunks, leaves etc.

Photosynthesis – The conversion of water and carbon-dioxide by plants into glucose and oxygen. Light is used as an energy source.

Plankton (Phyto and Zoo) – Floating organisms whose movements are more or less dependent on currents. Plant component is called phytoplankton the animal component-zooplankton.

Poikilotherms – Higher animals (mainly birds and mammals) that maintain their body temperature irrespective of the ambient temperature of the surroundings.

Pollutant – Extraneous substances present in concentrations that may harm organisms (humans, plants and animals) or exceed an environmental quality standard. The terms is frequently used synonymously with contaminant.

Population – A group of fish of one species which shares common ecological and genetic features. The stocks defined for the purpose of stock assessment and management do not necessarily coincide with self-contained populations.

Precautionary approach – Set of measures taken to implement the Precautionary Principle. A set of agreed cost-effective measures and actions, including future courses of action, which ensures prudent foresight, reduces or avoids risk to the resource, the environment, and the people to the extent possible, taking explicitly into account existing uncertainties and the potential consequences of being wrong.

Primary production – Synthesis of organic carbon by chlorophyll bearing organisms.

Primary producers – Plankton, periphyton, macro-vegetation and other communities that synthesize carbon.

Recruitment – Adding of individual into the fish stock through breeding and or stocking (artificial recruitment).

Renewable natural resource – Natural resources that, after exploitation, can return to their previous stock levels by natural processes of growth or replenishment.

Reservoir morphometry – Measurement of various dimensions of dam and reservoir that has bearing on productivity such as dam height, mean depth of the reservoir, shoreline length of reservoir, *etc.*

Riparian – Land adjacent to a stream/river.

Runoff – Portion of rainfall, melted snow or irrigation water that flows across the ground's surface and is eventually returned to streams.

Saprophobes – Relatively hardier organisms that multiply and occupy the niches vacated by saproxenes.

Saproxenes – Organisms that are extremely sensitive to habitat changes, which get eliminated first as organic pollution increases.

Scampi – The freshwater prawn, *Macrobrachium rosenbergii.*

Species enhancement – Inducting new species into the system to enhance productivity.

Species introduction – Inducting species outside its normal range of distribution.

Species diversity – The variety of species in a community, which can be expressed quantitatively in ways which reflect both the total numbers of species present and the extent to which the system is dominated by a small number of species.

Specific conductance – A measure of a material's ability to conduct an electric current.

Stakeholders - Individual/groups who have an interest or derive benefits from the system.

Stock enhancement - Enhanced capture fisheries. Augmenting stock by stocking to build a breeding population.

Sustainable use - The use of components of biological diversity in a way and at a rate that does not lead to the long-term decline of biological diversity, thereby maintaining its potential to meet the needs and aspirations of present and future generations.

Sustainable yield - The amount of biomass or the number of units that can be harvested currently in a fishery without compromising the ability of the populations/ecosystem to regenerate itself.

Thermocline (Thermal stratification) - Sudden decline in temperature at a particular depth in deep water bodies.

Trawling - A method of fishing that involves pulling a large fishing net through the water behind one or more boats.

Tropholytic zone - The dark region below the euphotic zone is called the tropholytic zone.

www.ingramcontent.com/pod-product-compliance
Ingram Content Group UK Ltd.
Pitfield, Milton Keynes, MK11 3LW, UK
UKHW021950270726
14060UKWH00002B/442

9 789388 173186